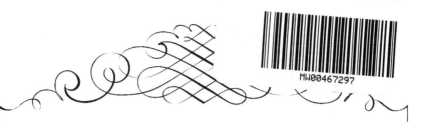

ISBN 978-1-332-59975-2
PIBN 10029019

This book is a reproduction of an important historical work. Forgotten Books uses state-of-the-art technology to digitally reconstruct the work, preserving the original format whilst repairing imperfections present in the aged copy. In rare cases, an imperfection in the original, such as a blemish or missing page, may be replicated in our edition. We do, however, repair the vast majority of imperfections successfully; any imperfections that remain are intentionally left to preserve the state of such historical works.

1 MONTH OF
FREE
READING

at

www.ForgottenBooks.com

By purchasing this book you are eligible for one month membership to ForgottenBooks.com, giving you unlimited access to our entire collection of over 700,000 titles via our web site and mobile apps.

To claim your free month visit:
www.forgottenbooks.com/free29019

English
Français
Deutsche
Italiano
Español
Português

www.forgottenbooks.com

Mythology Photography **Fiction**
Fishing Christianity **Art** Cooking
Essays Buddhism Freemasonry
Medicine **Biology** Music **Ancient**
Egypt Evolution Carpentry Physics
Dance Geology **Mathematics** Fitness
Shakespeare **Folklore** Yoga Marketing
Confidence Immortality Biographies
Poetry **Psychology** Witchcraft
Electronics Chemistry History **Law**
Accounting **Philosophy** Anthropology
Alchemy Drama Quantum Mechanics
Atheism Sexual Health **Ancient History**
Entrepreneurship Languages Sport
Paleontology Needlework Islam
Metaphysics Investment Archaeology
Parenting Statistics Criminology
Motivational

MATHEMATICS

FOR

THE PRACTICAL MAN

EXPLAINING SIMPLY AND QUICKLY
ALL THE ELEMENTS OF

ALGEBRA, GEOMETRY, TRIGONOMETRY,
LOGARITHMS, COÖRDINATE
GEOMETRY, CALCULUS
WITH ANSWERS TO PROBLEMS

BY

GEORGE HOWE, M.E.

ILLUSTRATED

SEVENTH THOUSAND

NEW YORK
D. VAN NOSTRAND COMPANY
25 PARK PLACE
1918

5

Stanhope Press
F. H. GILSON COMPANY
BOSTON, U.S.A.

DEDICATED TO

Brown Ayres, Ph.D.

PRESIDENT OF THE UNIVERSITY OF TENNESSEE
"MY GOOD FRIEND AND GUIDE."

PREFACE

In preparing this work the author has been prompted by many reasons, the most important of which are:

The dearth of short but complete books covering the fundamentals of mathematics.

The tendency of those elementary books which "begin at the beginning" to treat the subject in a popular rather than in a scientific manner.

Those who have had experience in lecturing to large bodies of men in night classes know that they are composed partly of practical engineers who have had considerable experience in the operation of machinery, but no scientific training whatsoever; partly of men who have devoted some time to study through correspondence schools and similar methods of instruction; partly of men who have had a good education in some non-technical field of work but, feeling a distinct calling to the engineering profession, have sought special training from night lecture courses; partly of commercial engineering salesmen, whose preparation has been non-technical and who realize in this fact a serious handicap whenever an important sale is to be negotiated and they are brought into competition with the skill of trained engineers; and finally, of young men leaving high schools and academies anxious to become engineers but who are unable to attend college for that purpose. Therefore it is apparent that with this wide difference in the degree of preparation of its students any course of study must begin with studies which are quite familiar to a large number but which have been forgotten or perhaps never undertaken by a large number of others.

And here lies the best hope of this textbook. It "begins at the beginning," assumes no mathematical knowledge beyond arithmetic on the part of the student, has endeavored to gather together in a concise and simple yet accurate and scientific form those fundamental notions of mathematics without which any studies in engineering are impossible, omitting the usual diffuseness of elementary works, and making no pretense at elaborate demonstrations, believing that where there is much chaff the seed is easily lost.

I have therefore made it the policy of this book that no technical difficulties will be waived, no obstacles circumscribed in the pursuit of any theory or any conception. Straightforward discussion has been adopted; where obstacles have been met, an attempt has been made to strike at their very roots, and proceed no further until they have been thoroughly unearthed.

With this introduction, I beg to submit this modest attempt to the engineering world, being amply repaid if, even in a small way, it may advance the general knowledge of mathematics.

GEORGE HOWE.

New York, *September*, 1910.

TABLE OF CONTENTS

MATHEMATICS

CHAPTER I

FUNDAMENTALS OF ALGEBRA

Addition and Subtraction

As an introduction to this chapter on the fundamental principles of algebra, I will say that it is absolutely essential to an understanding of engineering that the fundamental principles of algebra be thoroughly digested and redigested, — in short, literally soaked into one's mind and method of thought.

Algebra is a very simple science — extremely simple if looked at from a common-sense standpoint. If not seen thus, it can be made most intricate and, in fact, incomprehensible. It is arithmetic simplified, — a short cut to arithmetic. In arithmetic we would say, if one hat costs 5 cents, 10 hats cost 50 cents. In algebra we would say, if one a costs 5 cents, then 10 a cost 50 cents, a being used here to represent "hat." a is what we term in algebra a symbol, and all quantities are handled by means of such symbols. a is presumed to represent one thing; b, another symbol, is presumed to represent another thing, c another, d another, and so on for any number of objects. The usefulness and simplicity, therefore, of using symbols to represent objects is obvious. Suppose a merchant in the furniture business to be taking stock. He would go through his stock rooms and, seeing 10 chairs, he would actually write down "10 *chairs*"; 5 tables, he would actually write out "5 *tables*"; 4 beds, he would actually write this out, and so on. Now, if he had at the start agreed to represent chairs by the letter a, tables by the letter b, beds by the letter c, and so on, he would have been saved the necessity of writing down the names of these articles each time, and could have written 10 a, 5 b, and 4 c.

Definition of a Symbol.—A symbol is some letter by which it is agreed to represent some object or thing.

When a problem is to be worked in algebra, the first thing necessary is to make a choice of symbols, namely, to assign certain letters to each of the different objects concerned with the problem, — in other words, to get up a code. When this code is once established it must be rigorously maintained; that is, if, in the solution of any problem or set of problems, it is once stipulated that a shall represent a chair, then wherever a appears it means a chair, and wherever the word *chair* would be inserted an a must be placed—the code must not be changed.

Positivity and Negativity.—Now, in algebraic thought, not only do we use symbols to represent various objects and things, but we use the signs plus $(+)$ or minus $(-)$ before the symbols, to indicate what we call the *positivity* or *negativity* of the object.

Addition and Subtraction. — Algebraically, if, in going over his stock and accounts, a merchant finds that he has 4 tables in stock, and on glancing over his books finds that he owes 3 tables, he would represent the 4 tables in stock by such a form as $+4\,a$, a representing table; the 3 tables which he owes he would represent by $-3\,a$, the plus sign indicating that which he has on hand and the minus sign that which he owes. Grouping the quantities $+4\,a$ and $-3\,a$ together, in other words, striking a balance, one would get $+a$, which represents the one table which he owns over and above that which he owes. The plus sign, then, is taken to indicate all things on hand, all quantities greater than zero. The minus sign is taken to indicate all those things which are owed, all things less than zero.

Suppose the following to be the inventory of a certain quantity of stock: $+8\,a$, $-2\,a$, $+6\,b$, $-3\,c$, $+4\,a$, $-2\,b$, $-2\,c$, $+5\,c$. Now, on grouping these quantities together and striking a balance, it will be seen that there are 8 of those things which are represented by a on hand; likewise 4 more, represented by $4\,a$, on hand; 2 are owed, namely, $-2\,a$. Therefore, on grouping $+8\,a$, $+4\,a$, and $-2\,a$ together, $+10\,a$ will be the result. Now, collecting those terms representing the objects which we have called b, we have $+6\,b$ and $-2\,b$, giving as a result $+4\,b$. Grouping $3\,c$, $-2\,c$, and $+5\,c$ together will give o, because $+5\,c$ represents 5 c's on hand, and $-3\,c$ and $-2\,c$ represent that 5 c's are owed; therefore, these quantities neutralize and strike a balance. Therefore,

$$+8\,a-2\,a+6\,b-3\,c+4\,a-2\,b-2\,c+5\,c$$

reduces to $\qquad +10\,a+4\,b.$

This process of gathering together and simplifying a collection of terms having different signs is what we call in algebra *addition* and *subtraction*. Nothing is more simple, and yet nothing should be more thoroughly understood before proceeding further. It is obviously impossible to add one table to one chair and thereby get two chairs, or one book to one hat and get two books; whereas it is perfectly possible to add one book to another book and get two books, one chair to another chair and thereby get two chairs.

Rule. — *Like symbols can be added and subtracted, and only like symbols.*

$a+a$ will give $2\,a$; $3\,a+5\,a$ will give $8\,a$; $a+b$ will not give $2\,a$ or $2\,b$, but will simply give $a+b$, this being the simplest form in which the addition of these two terms can be expressed.

Coefficients. — In any term such as $+8\,a$ the plus sign indicates that the object is on hand or greater than zero, the 8 indicates the number of them on hand, it is the numerical part of the term and is called the *coefficient*, and the a indicates the nature of the object, whether it is a chair or a book or à table that we have represented by the symbol a. In the term $+6\,a$, the plus $(+)$ sign indicates that the object is owned, or greater than zero, the 6 indicates the number of objects on hand, and the a their nature. If a man has $20 in his pocket and he owes $50, it is evident that if he paid up as far as he could, he would still owe $30. If we had represented $1 by the letter a, then the $20 in his pocket would be represented by $+20\,a$, the $50 that he owed by $-50\,a$. On grouping these terms together, which is the same process as the settling of accounts, the result would be $-30\,a$.

Algebraic Expressions. — An algebraic expression consists of two or more terms; for instance, $+\,a+b$ is an algebraic expression; $+\,a+2\,b+c$ is an algebraic expression; $+3\,a+5\,b+6\,b+c$ is another algebraic expression, but this last one can be written more simply, for the $5\,b$ and $6\,b$ can be grouped together in one term, making $11\,b$, and the expression now becomes $+3\,a+11\,b+c$, which is as simple as it can be written. It is always advisable to group together into the smallest number of terms any algebraic expression wherever it is met in a problem, and thus simplify the manipulation or handling of it.

When there is no sign before the first term of an expression the plus (+) sign is intended.

To subtract one quantity from another, change the sign and then group the quantities into one term, as just explained. Thus: to subtract $4a$ from $+12a$ we write $-4a + 12a$, which simplifies into $+8a$. Again, subtracting $2a$ from $+6a$ we would have $-2a + 6a$, which equals $+4a$.

PROBLEMS

Simplify the following expressions:

1. $10a + 5b + 6c - 8a - 3d + b.$

2. $a - b + c - 10a - 7c + 2b.$

3. $10d + 3z + 8b - 4d - 6z - 12b + 5a - 3d$ $+ 8z - 10a + 8b - 5a - 6z + 10b.$

4. $5x - 4y + 3z - 2x + 4y + x + z + a - 7x$ $+ 6y.$

5. $3b - 2a + 5c + 7a - 10b - 8c + 4a - b + c.$

6. $-2x + a + b + 10y - 6x - y - 7a + 3b + 2y.$

7. $4x - y + z + x + 15z - 3x + 6y - 7y + 12z.$

CHAPTER II

FUNDAMENTALS OF ALGEBRA

Multiplication and Division

WE have seen how the use of algebra simplifies the operations of addition and subtraction, but in multiplication and division this simplification is far greater, and the great weapon of thought which algebra is to become to the student is now realized for the first time. If the student of arithmetic is asked to multiply one foot by one foot, his result is one square foot, the square foot being very different from the foot. Now, ask him to multiply one chair by one table. How can he express the result? What word can he use to signify the result? Is there any conception in his mind as to the appearance of the object which would be obtained by multiplying one chair by one table? In algebra all this is simplified. If we represent a table by a, and a chair by b, and we multiply a by b, we obtain the expression ab, which represents in its entirety the multiplication of a chair by a table. We need no word, no name by which to call it; we simply use the form ab, and that carries to our mind the notion of the thing which we call a multiplied by the thing which we call b. And thus the form is carried without any further thought being given to it.

Exponents. — The multiplication of a by a may be represented by aa. But here we have a further short cut, namely, a^2. This 2, called an *exponent*, indicates that two a's have been multiplied by each other; $a \times a \times a$ would give us a^3, the 3 indicating that three a's have been multiplied by one another; and so on. The exponent simply signifies the number of times the symbol has been multiplied by itself.

Now, suppose a^2 were multiplied by a^3, the result would be a^5, since a^2 signifies that 2 a's are multiplied together, and a^3 indicates that 3 a's are multiplied together; then multiplying these two expressions by each other simply indicates that 5 a's are multiplied together. $a^3 \times a^7$ would likewise give us a^{10}, $a^4 \times a^4$ would give us a^8, $a^4 \times a^4 \times a^2 \times a^3$ would give us a^{13}, and so on.

Rule. — *The multiplication by each other of symbols representing similar objects is accomplished by adding their exponents.*

Indentity of Symbols. — Now, in the foregoing it must be clearly seen that the combined symbol ab is different from either a or b; ab must be handled as differently from a or b as c would be handled; in other words, it is an absolutely new symbol. Likewise a^2 is as different from a as a square foot is from a linear foot, and a^3 is as different from a^2 as one cubic foot is from one square foot. a^2 is a distinct symbol. a^3 is a distinct symbol, and can only be grouped together with other a^3's. For example, if an algebraic expression such as this were met:

$$a^2 + a + ab + a^3 + 3 a^2 - 2 a - ab,$$

to simplify it we could group together the a^2 and the $+3 a^2$, giving $+4 a^2$; the $+a$ and the $-2 a$ give $-a$; the $+ab$ and the $-ab$ neutralize each other; there is only one term with the symbol a^3. Therefore the above expression simplified would be $4 a^2 - a + a^3$. This is as simple as it can be expressed. Above all things the most important is never to group unlike symbols together by addition and subtraction. Remember fundamentally that a, b, ab, a^2, a^3, a^4 are all separate and distinct symbols, each representing a separate and distinct thing.

Suppose we have $a \times b \times c$. It gives us the term abc. If we have $a^2 \times b$ we get a^2b. If we have $ab \times ab$, we get a^2b^2. If we have $2 ab \times 2 ab$ we get $4 a^2b^2$; $6 a^2b^3 \times 3 c$, we get $18 a^2b^3c$; and so on. Whenever two terms are multiplied by each other, the coefficients are multiplied together, and the similar parts of the symbols are multiplied together.

· **Division.** — Just as when in arithmetic we write down $\frac{2}{3}$ to mean 2 divided by 3, in algebra we write $\frac{a}{b}$ to mean a divided by b. a is called a numerator and b a denominator, and the expression $\frac{a}{b}$ is called a fraction.   If a^3 is multiplied by a^2, we have seen that the result is a^5, obtained by adding the exponents 3 and 2. If a^3 is divided by a^2, the result is a, which is obtained by subtracting 2 from 3.   Therefore $\dfrac{a^2b}{ab}$ would equal a, the a in the denominator dividing into a^2 in the numerator a times, and the b in the denominator canceling the b in the numerator.   Division is then simply the inverse of multiplication, which is patent.   On simplifying such an expression as $\dfrac{a^4b^2c^3}{a^2bc^5}$ we obtain $\dfrac{a^2b}{c^2}$, and so on.

Negative Exponents. — But there is a more scientific and logical way of explaining division as the inverse of multiplication, and it is thus: Suppose we have the fraction $\dfrac{1}{a^2}$.   This may be written a^{-2}, or the term b^2 may be written $\dfrac{1}{b^{-2}}$; that is, any term may be changed from the numerator of a fraction to the denominator by simply changing the sign of its exponent.   For example, $\dfrac{a^5}{a^2}$ may be written $a^5 \times a^{-2}$ Multiplying these two terms together, which is accomplished by adding their exponents, would give us a^3, 3 being the result of the addition of 5 and -2.   It is scarcely necessary, therefore, to make a separate law for division if one is made for multiplication, when it is seen that division simply changes the sign of the exponent.   This should be carefully considered and thought over by the pupil, for it is of great importance. Take such an expression as $\dfrac{a^2b^{-2}c^2}{abc^{-1}}$. Suppose all the symbols in the denominator are placed in the numerator, then we have $a^2b^{-2}c^2a^{-1}b^{-1}c$, or, simplifying, $ab^{-3}c^3$, which may be further written $\dfrac{ac^3}{b^3}$. The negative exponent is very serviceable, and it is well to become thoroughly familiar with it. The following examples should be worked by the student.

PROBLEMS

Simplify the following:

1. $2\,a \times 3\,b \times 3\,ab.$

2. $12\,a^2bc \times 4\,c^2b.$

3. $6\,x \times 5\,y \times 3\,xy.$

4. $4\,a^2bc \times 3\,abc \times 2\,a^5b \times 6\,b^2$

5. $\dfrac{a^2b^2c^3}{abc}.$

6. $\dfrac{a^4b^3c^2d}{a^2d^2}.$

7. $a^{-2} \times b^3 \times a^6b^2c.$

8. $abc^2 \times b\ ^2a\ ^1c^5 \times a^3b^3.$

9. $\dfrac{a^4b^{-6}c^3z}{a^2b^{-2}c}.$

10. $10\,a^2b \times 5\,a^{-1}bc^{-3} \times \dfrac{8\,ac^{-1}}{b^2a^{-4}} \times 10^{-1}a.$

11. $\dfrac{5\,a^2b^2c^2d^2}{45\,a^3 \times 6\,d^3}.$

CHAPTER III

FUNDAMENTALS OF ALGEBRA

Multiplication and Division (Continued).

HAVING illustrated and explained the principles of multiplication and division of algebraic terms, we will in this lecture inquire into the nature of these processes as they apply to algebraic expressions. Before doing this, however, let us investigate a little further into the principles of fractions.

Fractions. — We have said that the fraction $\frac{a}{b}$ indicated that a was divided by b, just as in arithmetic $\frac{1}{3}$ indicates that 1 is divided by 3. Suppose we multiply the fraction $\frac{1}{3}$ by 3, we obtain $\frac{3}{3}$, our procedure being to multiply the numerator 1 by 3. Similarly, if we had multiplied the fraction $\frac{a}{b}$ by 3, our result would have been $\frac{3a}{b}$.

Rule.— *The multiplication of a fraction by any quantity is accomplished by multiplying its numerator by that quantity;* thus, $\frac{2\,a^2}{b}$ multiplied by $3\,a$ would give $\frac{6\,a^3}{b}$. Conversely, when we divide a fraction by a quantity, we multiply its denominator by that quantity. Thus, the fraction $\frac{a}{b}$ when divided by $2b$ gives $\frac{a}{2b^2}$. Finally, should we multiply the numerator and the denominator by the same quantity, it is obvious that we do not change the value of the fraction, for we have multiplied and divided it by the same thing. From this it must not be deduced that adding the same quantity to both the numerator and the denominator of a fraction will not change its value. The beginner is likely to make this mistake, and he is here warned against it. Suppose we add to both the numerator and the denominator of the fraction $\frac{1}{3}$ the quantity 2. We will obtain $\frac{3}{5}$, which is different in value from $\frac{1}{3}$, proving that the addition or subtraction of the same quantity from both numerator and denominator of any fraction changes its value. The multiplication or division of both the numerator and the denominator by the same quantity does not alter the value of a fraction one whit.

Multiplying two fractions by each other is accomplished by multiplying their numerators together and multiplying their denominators together. Thus, $\frac{a}{b} \times \frac{d}{c}$ would give us $\frac{ad}{bc}$.

Suppose it is desired to add the fraction $\frac{1}{2}$ to the fraction $\frac{1}{3}$. Arithmetic teaches us that it is first necessary to reduce both fractions to a common denominator, which in this case is 6, viz.: $\frac{3}{6} + \frac{2}{6} = \frac{5}{6}$, the numerators being added if the denominators are of a common value. Likewise, if it is desired to add $\frac{a}{b}$ to $\frac{c}{d}$, we must reduce both of these fractions to a common denominator, which in this case is bd. (The common denominator of several denominators is a quantity into which any one of these denominators may be divided; thus b will divide into bd, d times, and d will divide into bd, b times.) Our fractions then become $\frac{ad}{bd} + \frac{cb}{bd}$. The denominators now having a common value, the fractions may be added by adding the numerators, resulting in $\frac{ad + cb}{bd}$. Likewise, adding the fractions $\frac{a}{3} + \frac{b}{2a} + \frac{c}{3a}$, we find that the common denominator in this case is $6a$. The first fraction becomes $\frac{2a^2}{6a}$, the second $\frac{3b}{6a}$ and the third $\frac{2c}{6a}$, the result being the fraction $\frac{2a^2 + 3b + 2c}{6a}$. This process will be taken up and explained in more detail later, but the student should make an attempt to apprehend the principles here stated and solve the problems given at the end of this lecture.

Law of Signs. — Like signs multiplied or divided give $+$ and unlike signs give . Thus:

$$+ 3a \times + 2a \text{ gives } + 6a^2,$$
$$\text{also} - 3a \times - 2a \text{ gives } + 6a^2,$$

while $+ 3\,a \times - 2\,a$ gives $- 6\,a^2$

or $- 3\,a \times + 2\,a$ gives $- 6\,a^2$;

furthermore $+ 8\,a^2$ divided by $+ 2\,a$ gives $+ 4\,a$,

and $- 8\,a^2$ divided by $- 2\,a$ gives $+ 4\,a$

while $- 8\,a^2$ divided by $+ 2\,a$ gives $- 4\,a$

or $+ 8\,a^2$ divided by $- 2\,a$ gives $- 4\,a$.

Multiplication of an Algebraic Expression by a Quantity. — As previously said, an algebraic expression consists of two or more terms. $3\,a$, $5\,b$, are terms, but $3\,a + 5\,b$ is an algebraic expression. If the stock of a merchant consists of 10 tables and 5 chairs, and he doubles his stock, it is evident that he must double the number of tables and also the number of chairs, namely, increase it to 20 tables and 10 chairs. Likewise, when an algebraic expression which consists of $3\,a + 2\,b$ is doubled, or, what is the same thing, multiplied by 2, each term must be doubled or multiplied by 2, resulting in the expression $6\,a + 4\,b$. Similarly, when an algebraic expression consisting of several terms is divided by any number, each term must be divided by that number.

Rule. — An algebraic expression must be treated as a unit. Whenever it is multiplied or divided by any quantity, each term of the expression must be multiplied or divided by that quantity. For example: Multiplying the expression $4x + 3y + 5xy$ by the quantity $3x$ will give the following result: $12x^2 + 9xy + 15x^2y$, obtained by multiplying each one of the separate terms by $3x$ successively.

Division of an Algebraic Expression by a Quantity. — Dividing the expression $6a^3 + 2a^2b + 4b^2$ by $2ab$ would result in the expression $\dfrac{3a^2}{b} + a + \dfrac{2b}{a}$, obtained by dividing each term successively by $2b$. This rule must be remembered, as its importance cannot be over-estimated. The numerator or denominator of a fraction consisting of one or two or more terms must be handled as a unit, this being one of the most important applications of this rule. For example, in the fraction $\dfrac{a+b}{a}$ or $\dfrac{a}{a+b}$, it is impossible to cancel out the a in the numerator and denominator, for the reason that if the numerator is divided by a, each term must be divided by a, and the operation upon the one term a without the same operation upon the term b would be erroneous. If the fraction $\dfrac{a+b}{a}$ is multiplied by 3, it becomes $\dfrac{3a+3b}{a}$ If the fraction $\dfrac{a-b}{a+b}$ is multiplied by $\dfrac{2}{3}$ it becomes $\dfrac{2a-2b}{3a+3b}$; and so on. Never forget that the numerator (or denominator) of a fraction consisting of two or more terms is an algebraic expression and must be handled as a unit.

Multiplication of One Algebraic Expression by Another — It is frequently desired to multiply an algebraic expression not only by a single term but by another algebraic expression consisting of two or more terms, in which case the first expression is multiplied throughout by each term of the second expression. The terms which result from this operation are then collected together by addition and subtraction and the result expressed in the simplest manner possible. Suppose it were desired to multiply $a + b$ by $c + d$. We would first multiply $a + b$ by c, which would give us $ac + bc$. Then we would multiply $a + b$ by d, which would give us $ad + bd$. Now, collecting the result of these two multiplications together, we obtain $ac + bc + ad + bd$, viz.:

$$\begin{array}{l} a + b \\ \underline{c + d} \\ ac + bc \\ \quad\quad\quad \underline{ad + bd} \\ ac + bc + ad + bd \end{array}$$

Again, let us multiply

$$2a + b - 3c$$

by

$$\underline{a + 2b - c}$$

$$2a^2 + ab - 3ac$$
$$\quad\quad 4ab \quad\quad\quad\quad + 2b^2 \quad 6bc$$
$$\quad\quad\quad\quad \underline{- 2ac \quad\quad\quad\quad - \ bc + 3c^2}$$

and we have

$$2a^2 + 5ab - 5ac + 2b^2 - 7bc + 3c^2.$$

The Division of one Algebraic Expression by Another. — This is somewhat more difficult to explain and understand than the foregoing. In general it may be said that, suppose we are required to divide the expression $6 a^2 + 17 ab + 12 b^2$ by $3 a + 4 b$, we would arrange the expression in the following way ·

$$
\begin{array}{l|l}
6 a^2 + 17\,ab + 12\,b^2 & 3 a + 4 b \\
\underline{6 a^2 + 8\,ab} & 2 a + 3 b \\
9\,ab + 12\,b^2 \\
\underline{9\,ab + 12\,b^2}
\end{array}
$$

$3 a$ will divide into $6 a^2$, $2 a$ times, and this is placed in the quotient as shown. This $2 a$ is then multiplied successively into each of the terms in the divisor, namely, $3 a + 4 b$, and the result, namely, $6 a^2 + 8 ab$, is placed beneath the dividend, as shown. A line is then drawn and this quantity subtracted from the dividend, leaving $9 ab$. The $+12 b^2$ in the dividend is now carried. Again, we observe that $3 a$ in the divisor will divide into $9 ab$, $+3 b$ times, and we place this term in the divisor. Multiplying $3 b$ by each of the terms of the divisor, as before, will give us $9 ab + 12 b^2$; and, subtracting this as shown, nothing remains, the final result of the division then being the expression $2 a + 3 b$.

This process should be studied and thoroughly understood by the student.

PROBLEMS

Solve the following problems:

1. Multiply the fraction $\dfrac{3\,a^2b^3c}{4\,x^2}$ by the quantity $3\,x$.

2. Divide the fraction $\dfrac{abc}{6\,d}$ by the quantity $3\,a$.

3. Multiply the fraction $\dfrac{a^2b^2c^2}{xy^3}$ by the fraction $\dfrac{a^3b^3}{6\,a}$ by the fraction $\dfrac{x^2y}{b}$

4. Multiply the expression $4\,x + 3\,y + 2\,z$ by the quantity $5\,x$.

5. Divide the expression $8\,a^2b + 4\,a^3b^3 - 2\,ab^2$ by the quantity $2\,ab$.

6. Multiply the expression $a + b$ by the expression $a - b$.

7. Multiply the expression $2\,a + b - c$ by the expression $3\,a - 2\,b + 4\,c$.

8. Divide the expression $a^2 - 2\,ab + b^2$ by $a - b$.

9. Divide the expression $a^3 + 3\,a^2b + 3\,ab^2 + b^3$ by $a + b$.

10. Multiply the fraction $\dfrac{a + b}{a - b}$ by $\dfrac{a - b}{a - b}$.

11. Multiply the fraction $\dfrac{3\,a}{c + d}$ by $\dfrac{c - d}{2}$ by $\dfrac{a + c}{a - c}$.

12. Multiply the fraction $\dfrac{a^{-2}bc^3}{4}$ by $\dfrac{b}{3\,a^{-2}}$ by $\dfrac{a}{b}$.

13. Add together the fractions $\dfrac{2\,a}{b} + \dfrac{b}{4} + \dfrac{c}{b}$.

14. Add together the fractions $\dfrac{2}{3\,a^2} - \dfrac{4}{2\,a} + \dfrac{c}{6}$.

15. Add together the fractions $\dfrac{10\,a^2}{b} + \dfrac{b}{4\,b} - \dfrac{x}{2\,c} + \dfrac{d}{6}$.

16. Add together the fractions $\dfrac{a+b}{2\,a} + \dfrac{b-c}{4\,b}$.

17. Add together the fractions $\dfrac{a}{a+b} - \dfrac{2}{5\,a}$.

Factoring

Definition of a Factor. — A factor of a quantity is one of the two or more parts which when multiplied together give the quantity. A factor is an integral part of a quantity, and the ability to divide and sub-divide a quantity, be it a single term or a whole expression, into those factors whose multiplication has created it, is very valuable.

Factoring. — Suppose we take the number 6. Its factors are readily detected as 2 and 3. Likewise the factors of 10 are 5 and 2. The factors of 18 are 9 and 2; or, better still, $3 \times 3 \times 2$. The factors of 30 are $3 \times 2 \times 5$; and so on. The factors of the algebraic expression ab are readily detected as a and b, because their multiplication created the term ab. The factors of $6\,abc$ are 3, 2, a, b and c. The factors of $25\,ab$ are 5, 5, a and b, which are quite readily detected.

The factors of an expression consisting of two or more terms, however, are not so readily seen and sometimes require considerable ingenuity for their detection. Suppose we have an algebraic expression in which all of the terms have one or more common factors, — that is, that one or more like factors appear in the make-up of each term. It is often desirable in this case to remove the common factors from the several terms, and in order to do this without changing the value of any of the terms, the common factor or factors are placed outside of a parenthesis and the terms from which they have been removed placed within the parenthesis in their simplified form. Thus, in the algebraic expression $6\,a^2b + 3\,a^3$, $3\,a^2$ is a common factor of both terms; therefore we may write the expression, without changing its value, in the following manner: $3\,a^2\,(2\,b + a)$. The term $3\,a^2$ written outside of the parenthesis indicates that it must be multiplied into each of the separate terms within the parenthesis. Likewise, in the expression $12\,xy + 4\,x^3 + 6\,x^2z + 8\,xz$, $2\,x$ is a common factor of each of the terms, and the expression may be written $2\,x\,(6\,y + 2\,x^2 + 3\,xz + 4\,z)$. It is often desirable to factor in this simple manner.

Still further suppose we have $a^2 + ab + ac + bc$; we can take a out of the first two terms and c out of the last two, thus: $a\,(a + b) + c\,(a + b)$. Now we have two separate terms and taking $(a + b)$ out of each we have $(a + b) \times (a + c)$. Likewise, in the expression

$$6\,x^2 + 4\,xy - 3\,zx - 2\,zy$$

we have

$$2\,x\,(3\,x + 2\,y) - z\,(3\,x + 2\,y),$$

or,

$$(3\,x + 2\,y) \times (2\,x - z).$$

Now, suppose we have the expression $a^2 - 2\,ab + b^2$. We readily detect that this quantity is the result of multiplying $a - b$ by $a - b$; the first and last terms are respectively the squares of a and b, while the middle term is equal to twice the product of a and b. Any expression where this is the case is a perfect square; thus, $9\,x^2 - 12\,xy + 4\,y^2$ is the square of $3\,x - 2\,y$, and may be written $(3\,x - 2\,y)^2$. Remembering these facts, a perfect square is readily detected.

The product of the sum and difference of two terms such as $(a + b) \times (a - b)$ equals $a^2 - b^2$; or, briefly, the product of the sum and difference of two numbers is equal to the difference of their squares.

By trial it is often easy to discover the factors of algebraic expressions; for example, $2\,a^2 + 7\,ab + 3\,b^2$ is readily detected to be the product of $2\,a + b$ and $a + 3\,b$.

Factor the following:

1. $30\,a^2b.$

2. $48\,a^4c.$

3. $30\,x^2y^4z^3.$

4. $144\,x^2a^2.$

5. $\dfrac{12\,ab^2c^3}{4\,a^2b^2}.$

6. $\dfrac{10\,xy^2}{2\,x^2y}.$

7. $2\,a^2 + ab - 2\,ac - bc.$

8. $3 x^2 + xy + 3 xc + cy.$

9. $2 x^2 + 5 xy + 2 xz + 5 yz.$

10. $a^2 - 2 ab + b^2.$

11. $4 x^2 - 12 xy + 9 y^2.$

12. $81 a^2 + 90 ab + 25 b^2.$

13. $16 c^2 - 48 ca + 36 a^2.$

14. $4 x^3 y + 5 xzy^2 - 10 xzy.$

15. $30 ab + 15 abc - 5 bc.$

16. $81 x^2 y^2 \quad 25 a^2.$

17. $a^4 - 16 b^4.$

18. $144 x^4 y^2 - 64 z^2.$

19. $4 a^2 \quad 8 ac + 4 c.$

20. $16 y^2 + 8 xy + x^2.$

21. $6 y^2 - 5 xy \quad 6 x^2.$

22. $4 a^2 - 3 ab - 10 b^2.$

23. $6 y^2 - 13 xy + 6 x^2.$

24. $2 a^2 - 5 ab - 3 b^2.$

25. $2 a^2 + 9 ab + 10 b^2.$

CHAPTER V

FUNDAMENTALS OF ALGEBRA

Involution and Evolution

WE have in a previous chapter discussed the process by which we can raise an algebraic term and even a whole algebraic expression to any power desired, by multiplying it by itself. Let us now investigate the method of finding the square root and the cube root of an algebraic expression, as we are frequently called upon to do.

The square root of any term such as a^2, a^4, a^6, and so on, will be, respectively, $\pm a$, $\pm a^2$, and $\pm a^3$, obtained by dividing the exponents by 2. The plus-or-minus sign $(\pm)$ shows that either $+a$ or $-a$ when squared would give us $+a^2$. On taking the square root, therefore, the plus-or-minus sign $(\pm)$ is always placed before the root. This is not the case in the cube root, however. Likewise, the cube root of such terms as a^3, a^6, a^9, and so on, would be respectively a, a^2 and a^3, obtained by dividing the exponents by 3. Similarly, the square root of $4\,a^4b^6$ will be seen to be $\pm\,2\,a^2b^3$, obtained by taking the square root of each factor of the term. And likewise the cube root of $-27\,a^9b^6$ will be $-3\,a^3b^2$. These facts are so self-evident that it is scarcely necessary to dwell upon them. However, the detection of the square and the cube root of an algebraic expression consisting of several terms is by no means so simple.

Square Root of an Algebraic Expression. — Suppose we multiply the expression $a + b$ by itself. We obtain $a^2 + 2\,ab + b^2$. This is evidently the square of $a + b$. Suppose then we are given this expression and asked to determine its square root. We proceed in this manner: Take the square root of the first term and isolate it, calling it the trial root. The square root of a^2 is a; therefore place a in the trial root. Now square a and subtract this from the original expression, and we have the remainder $2\,ab + b^2$. For our trial divisor we proceed as follows: Double the part of the root already found, namely, a. This gives us $2\,a$. $2\,a$ will go into $2\,ab$, the first term of the remainder, b times. Add b to the trial root, and the same becomes $a + b$. Now multiply the trial divisor by b, it gives us $2\,ab + b^2$, and subtracting this from our former remainder, we have nothing left. The square root of our expression, then, is seen to be $a + b$, viz.:

$$
\begin{array}{r|l}
a^2 + 2\,ab + b^2 & a + b \\
a^2 & \\
\hline
2\,a + b \quad\, & 2\,ab + b^2 \\
& 2\,ab + b^2 \\
\hline
\end{array}
$$

Likewise we see that the square root of $4\,a^2 + 12\,ab + 9\,b^2$ is $2\,a + 3\,b$, viz.:

$$
\begin{array}{ll}
4\,a^2 + 12\,ab + 9\,b^2 & \underline{\;2\,a + 3\,b\;} \\
4\,a^2 & \\
\end{array}
$$

$$
4\,a + 3\,b \;\Big|\; \begin{array}{l} 12\,ab + 9\,b^2 \\ 12\,ab + 9\,b^2 \end{array}
$$

The Cube Root of an Algebraic Expression. — If we multiply $a + b$ by itself three times, in other words, cube the expression, we obtain $a^3 + 3\,a^2b + 3\,ab^2 + b^3$. It is evident, therefore, that if we had been given this latter expression and asked to find its cube root, our result should be $a + b$. In finding the cube root, $a + b$, we proceed thus: We take the cube root of the first term, namely, a, and place this in our trial root. Now cube a, subtract the a^3 thus obtained from the original expression, and we have as a remainder $3\,a^2b + 3\,ab^2 + b^3$ Now our trial divisor will consist as follows: Square the part of the root already found and multiply same by 3. This gives us $3\,a^2$. Divide $3\,a^2$ into the first term of the remainder, namely, $3\,a^2b$, and it will go b times. b then becomes the second term of the root. Now add to the trial divisor three times the first term of the root multiplied by the second term of the root, which gives us $3\,ab$. Then add the second term of the root square, namely, b^2. Our full divisor now becomes $3\,a^2 + 3\,ab + b^2$. Now multiply this full divisor by b and subtract this from the former remainder, namely,

$3\,a^2b + 3\,ab^2 + b^3$, and, having nothing left, we see that the cube root of our original expression is $a + b$, viz.:

$$a^3 + 3\,a^2b + 3\,ab^2 + b^3 \,\big|\, \underline{a + b}$$
$$a^3$$

$3\,a^2 + 3\,ab + b^2$ $\big|\, 3\,a^2b + 3\,ab^2 + b^3$
$ 3\,a^2b + 3\,ab^2 + b^3$

Likewise the cube root of $27\,x^3 + 27\,x^2 + 9\,x + 1$ is seen to be $3\,x + 1$, viz.:

$$27\,x^3 + 27\,x^2 + 9\,x + 1 \,\big|\, \underline{3\,x + 1}$$
$$27\,x^3$$

$27\,x^2 + 9\,x + 1$ $\big|\, 27\,x^2 + 9\,x + 1$
$ 27\,x^2 + 9\,x + 1$

Find the square root of the following expressions:

1. $16\,x^2 + 24\,xy + 9\,y^2$.
2. $4\,a^2 + 4\,ab + b^2$.
3. $36\,x^2 + 24\,xy + 4\,y^2$.
4. $25\,a^2 - 20\,ab + 4\,b^2$
5. $a^2 + 2\,ab + 2\,ac + 2\,bc + b^2 + c^2$.

Find the cube root of the following expressions:

1. $8\,x^3 + 36\,x^2y + 54\,xy^2 + 27\,y^3$.
2. $x^3 + 6\,x^2y + 12\,xy^2 + 8\,y^3$.
3. $27\,a^3 + 81\,a^2b + 81\,ab^2 + 27\,b^3$.

CHAPTER VI

FUNDAMENTALS OF ALGEBRA

Simple Equations

AN equation is the expression of the equality of two things; thus, $a = b$ signifies that whatever we call a is equal to whatever we call b; for example, one pile of money containing $100 in one shape or another is equal to any other pile containing $100. It is evident that if a quantity is added to or subtracted from one side of an equation or equality, it must be added to or subtracted from the other side of the equation or equality, in order to retain the equality of the two sides; thus, if $a = b$, then $a + c = b + c$ and $a - c = b - c$. Similarly, if one side of an equation is multiplied or divided by any quantity, the other side must be multiplied or divided by the same quantity; thus,

if $\qquad\qquad a = b,$

then $\qquad\qquad ac = bc$

and $\qquad\qquad \dfrac{a}{c} = \dfrac{b}{c}.$

Similarly, if one side of an equation is squared, the other side of the equation must be squared in order to retain the equality. In general, whatever is done to one side of an equation must also be done to the other side in order to retain the equality of both sides. The logic of this is self-evident.

Transposition. — Suppose we have the equation $a + b = c$. Subtract b from both sides, and we have $a + b - b = c - b$. On the left-hand side of the equation the $+b$ and the $-b$ will cancel out, leaving a, and we have the result $a = c - b$. Compare this with our original equation, and we will see that they are exactly alike except for the fact that in the one b is on the left-hand side of the equation, in the other b is on the right-hand side of the equation; in one case its sign is plus, in the other case its sign is minus. This indicates that in order to change a term from one side of an equation to the other side it is simply necessary to change its sign; thus,

$$a - c + b = d$$

may be transposed into the equation

$$a = c - b + d,$$

or into the form $a - d = c - b$,

or into the form $-d = c - a - b$.

Any term may be transposed from one side of an equation to the other simply by changing its sign.

Adding or Subtracting Two Equations. — When two equations are to be added to one another their corresponding sides are added to one another; thus, $a + c = b$ when added to $a = d + e$ will give $2a + c = b + d + e$. Likewise $3a + b = 2c$ when subtracted from $10a + 2b = 6c$ will yield $7a + b = 4c$.

Multiplying or Dividing Two Equations by one Another. — When two equations are multiplied or divided by one another their corresponding sides must be multiplied or divided by one another; thus, $a = b$ multiplied by $c = d$ will give $ac = bd$, also $a = b$ divided by $c = d$ will give $\dfrac{a}{c} = \dfrac{b}{d}$

Solution of an Equation. — Suppose we have such an equation as $4x + 10 = 2x + 24$, and it is desired that this equation be solved for the value of x; that is, that the value of the unknown quantity x be found. In order to do this, the first process must always be to group the terms containing x on one side of the equation by themselves and all the other terms in the equation on the other side of the equation. In this case, grouping the terms containing the unknown quantity x on the left-hand side of the equation we have

$$4x - 2x = 24 - 10.$$

Now, collecting the like terms, this becomes

$$2x = 14.$$

The next step is to divide the equation through by the coefficient of x, namely, 2. Dividing the left-hand side by 2, we have x. Dividing the right-hand side by 2, we have 7. Our equation, therefore, has resolved itself into

$$x = 7.$$

We therefore have the value of x. Substituting this value in the original equation, namely,

$$4x + 10 = 2x + 24,$$

we see that the equation becomes

$$28 + 10 = 14 + 24,$$

or, $$38 = 38,$$

which proves the result.

The process above described is the general method of solving for an unknown quantity in a simple equation.

Let us now take the equation

$$2cx + c = 40 - 5x.$$

This equation contains two unknown quantities, namely, c and x, either of which we may solve for. x is usually, however, chosen to represent the unknown quantity, whose value we wish to find, in an algebraic expression; in fact, x, y and z are generally chosen to represent unknown quantities. Let us solve for x in the above equation. Again we group the two terms containing x on one side of the equation by themselves and all other terms on the other side, and we have

$$2cx + 5x = 40 - c.$$

On the left-hand side of the equation we have two terms containing x as a factor. Let us factor this expression and we have

$$x(2c + 5) = 40 - c.$$

Dividing through by the coefficient of x, which is the parenthesis in this case, just as simple a coefficient to handle as any other, and we have

$$x = \frac{40 - c}{2c + 5}.$$

This final result is the complete solution of the equation as to the value of x, for we have x isolated on one side of the equation by itself, and its value on the other side. *In any equation containing any number of unknown quantities represented by symbols, the complete solution for the value of any one of the unknowns is accomplished when we have isolated this unknown on one side of the equation by itself. This is, therefore, the whole object of our solution.*

It is true that the value of x above shown still contains an unknown quantity, c. Suppose the numerical value of c were now given, we could immediately find the corresponding numerical value of x; thus, suppose c were equal to 2, we would have

$$x = \frac{40 - 2}{4 + 5},$$

or,
$$x = \frac{38}{9}.$$

This is the numerical value of x, corresponding to the numerical value 2 of c. If 4 had been assigned as the numerical value of c we should have

$$x = \frac{40 - 4}{8 + 5} = \frac{36}{13}.$$

Clearing of Fractions. — The above simple equations contained no fractions. Suppose, however, that we are asked to solve the equation

$$\frac{x}{4} + \frac{6}{2} = \frac{3\,x}{2} + \frac{5}{6}.$$

Manifestly this equation cannot be treated at once in the manner of the preceding example. The first step in solving such an equation is the removal of all the denominators of the fractions in the equation, this step being called the *Clearing of Fractions.*

As previously seen, in order to add together the fractions $\frac{1}{2}$ and $\frac{1}{3}$ we must reduce them to a common denominator, 6. We then have $\frac{3}{6} + \frac{2}{6} = \frac{5}{6}$. Likewise, in equations, before we can group or operate upon any one of the terms we must reduce them to a common denominator. The common denominator of several denominators is any number into which any one of the various denominators will divide, and the *least common denominator* is the smallest such number. The product of all the denominators — that is, multiplying them all together — will always give a *common denominator*, but not always the *least common denominator*. The *least common denominator*, being the smallest common denominator, is always desirable in preference to a larger number; but some ingenuity is needed frequently in detecting it. The old rule of withdrawing all factors common to at least two denominators and multiplying them together, and then by what is left of the denominators, is probably the easiest and simplest way to proceed. Thus, suppose we have the denominators 6, 8, 9 and 4. 3 is common to both 6 and 9, leaving respectively 2 and 3. 2 is common to 2, 8 and 4, leaving respectively 1, 4 and 2, and still further common to 4 and 2. Finally, we have removed the common factors 3, 2 and 2, and we have left in the denominators 1, 2, 3 and 1. Multiplying all of these together we have 72, which is the Least Common Denominator of these numbers, viz.:

$$
\begin{array}{r|l}
3 & 6, 8, 9, 4 \\
2 & 2, 8, 3, 4 \\
2 & 1, 4, 3, 2 \\
\hline
 & 1, 2, 3, 1
\end{array}
$$

$$3 \times 2 \times 2 \times 1 \times 2 \times 3 \times 1 = 72.$$

Having determined the Least Common Denominator, or any common denominator for that matter, the next step is to multiply each denominator by such a quantity as will change it into the Least Common Denominator. If the denominator of a fraction is multiplied by any quantity, as we have previously seen, the numerator must be multiplied by that same quantity, or the value of the fraction is changed. Therefore, in multiplying the denominator of each fraction by a quantity, we must also multiply the numerator. Returning to the equation which we had at the outset, namely, $\frac{x}{4} + \frac{6}{2} = \frac{3x}{2} + \frac{5}{6}$, we see that the common denominator here is

12. Our equation then becomes $\frac{3\,x}{12} + \frac{36}{12} = \frac{18x}{12} + \frac{10}{12}$. We have previously seen that the multiplication or division of both sides of an equation by the same quantity does not alter the value of the equation. Therefore we can at once multiply both sides of this equation by 12. Doing so, all the denominators disappear. This is equivalent to merely canceling all the denominators, and the equation is now changed to the simple form $3\,x + 36 = 18\,x + 10$. On transposition this becomes

$$3\,x - 18\,x = 10 - 36,$$

or
$$15\,x = -26,$$

or
$$x = \frac{26}{15},$$

or
$$+\,x = \frac{+26}{15}.$$

Again, let us now take the equation

$$\frac{2\,x}{5\,c} + \frac{10}{c^2} = \frac{x}{3}.$$

The *least common denominator* will at once be seen to be $15\,c^2$. Reducing all fractions to this common denominator we have

$$\frac{6\,cx}{15\,c^2} + \frac{150}{15\,c^2} - \frac{5\,c^2x}{15\,c^2}.$$

Canceling all denominators, we then have

$$6\,cx + 150 = 5\,c^2x.$$

Transposing, we have

$$6\,cx - 5\,c^2x = -150.$$

Taking x as a common factor out of both of the terms in which it appears, we have

$$x\,(6\,c - 5\,c^2) = -150.$$

Dividing through by the parenthesis, we have

$$x = \frac{-150}{6\,c - 5\,c^2}.$$

This is the value of x. If some numerical value is given to c, such as 2, for instance, we can then find the corresponding numerical value of x by substituting the numerical value of c in the above, and we have

$$x = \frac{-150}{12 - 20} = \frac{-150}{-8} = 18.75.$$

In this same manner all equations in which fractions appear are solved.

PROBLEMS

Suppose we wish to make use of algebra in the solution of a simple problem usually worked arithmetically, taking, for example, such a problem as this: A man purchases a hat and coat for $15.00, and the coat costs twice as much as the hat. How much did the hat cost? We would proceed as follows: Let x equal the cost of the hat. Since the coat cost twice as much as the hat, then $2x$ equals the cost of the coat, and $x + 2x = 15$ is the equation representing the fact that the cost of the coat plus the cost of the hat equals $15; therefore, $3x = \$15$, from which $x = \$5$; namely, the cost of the hat was $5. $2x$ then equals $10, the cost of the coat. Thus many problems may be attacked.

Solve the following equations:

1. $6x - 10 + 4x + 3 = 2x + 20 - x + 15.$

2. $x + 5 + 3x + 6 = -10x + 25 + 8x.$

3. $cx + 4 + x = cx + 8.$ Find the numerical value of x if $c = 3$.

4. $\dfrac{x}{5} + 3 = \dfrac{8x}{2} + 4.$

5. $\dfrac{4x}{3} + \dfrac{3x}{5} + \dfrac{7}{2} = \dfrac{11}{3} + x.$

6. $\dfrac{x}{c} + \dfrac{10}{4c} = \dfrac{x}{3} + \dfrac{x}{12c}.$ Find the numerical value of x if $c = 3$.

7. $\dfrac{10\,c}{3} - \dfrac{cx}{c} + \dfrac{8}{5\,c} = \dfrac{3\,cx}{10} + \dfrac{15}{2\,c}$ Find the numerical value of x if $c = 6$.

8. $\dfrac{x}{a+b} - 2 + \dfrac{y}{3} = 1$.

9. $\dfrac{2\,x}{a} + 3\,x + \dfrac{2}{a-b} = x - \dfrac{3}{a^2}$.

10. $\dfrac{x}{a+b} + \dfrac{x}{a-b} = 10$.

11. Multiply $ax + b = cx - b$ by $2\,a - x = c + 10$.

12. Multiply $\dfrac{a}{3} + b = \dfrac{c}{d}$ by $x = y + 3$.

13. Divide $a^2 - b^2 = c$ by $a + b = c + 3$.

14. Divide $2\,a = 10\,y$ by $a = y + 2$.

15. Add $2\,a + 10 = x + 3 - d$ to $3\,a - 7 = 2\,d$.

16. Add $4\,ax + 2\,y = -10\,x$ to $2\,ax - 7\,y = 5$.

17. Add $15\,z^2 + x = 5$ to $3\,x = -10\,y + 7$.

18. Subtract $2\,a - d = 8$ from $8\,a + d = 12$.

19. Subtract $3\,x + 7 = 15\,x^2 + y$ from $6\,x + 5 = 18\,x^2$.

20. Subtract $\dfrac{2\,x}{10\,y} + c = 7$ from $\dfrac{10\,x}{5\,y} = 18$.

21. Multiply $\dfrac{x}{3\,a+b} - \dfrac{x}{3} - c$ by $\dfrac{x}{c-d} = \dfrac{2\,a+b}{6}$.

22. Solve the equation $\dfrac{1}{x} = -\dfrac{1}{x+1}$.

23. If a coat cost one-half as much as a gun and twice as much as a hat, and all cost together $100, what is the cost of each?

24. The value of a horse is $15 more than twice the value of a carriage, and the cost of both is $1000; what is the cost of each ?

25. One-third of Anne's age is 5 years less than one-half plus 2 years; what is her age?

26. A merchant has 10 more chairs than tables in stock. He sells four of each and adding up stock finds that he now has twice as many chairs as tables. How many of each did he have at first?

CHAPTER VII

FUNDAMENTALS OF ALGEBRA

Simultaneous Equations

As seen in the previous chapter, when we have a simple equation in which only one unknown quantity appears, such, for instance, as x, we can, by algebraic processes, at once determine the numerical value of this unknown quantity. Should another unknown quantity, such as c, appear in this equation, in order to determine the value of x some definite value must be assigned to c. However, this is not always possible. An equation containing two unknown quanti-. ties represents some manner of relation between these quantities. If two separate and distinct equations representing two separate and distinct relations which exist between the two unknown quantities can be found, then the numerical values of the unknown quantities become fixed, and either one can be determined without knowing the corresponding value of the other. The two separate equations are called simultaneous equations, since they represent simultaneous relations between the unknown quantity. The following is an example:

$$x + y = 10.$$
$$x - y = 4.$$

The first equation represents one relation between x and y. The second equation represents another relation subsisting between x and y. The solution for *the numerical value of x*, or that of y, from these two equations, consists in *eliminating* one of the unknowns, x or y as the case may be, by adding or subtracting, dividing or multiplying the equations by each other, as will be seen in the following. Let us now find the value of x in the first equation, and we see that this is

$$x = 10 - y.$$

Likewise in the second equation we have

$$x = 4 + y.$$

·These two values of x may now be equated (things equal to the same thing must be equal to each other), and we have

$$10 - y = 4 + y,$$

or,
$$2y = 4 - 10,$$
$$2y = -6$$
$$+ 2y = +6,$$
$$y = 3.$$

Now, this is the value of y. In order to find the value of x, we substitute this numerical value of y in one of the equations containing both x and y, such as the first equation, $x + y = 10$. Substituting, we have

$$x + 3 = 10.$$

Transposing, $\qquad x = 10 - 3,$

$$x = 7.$$

Here, then, we have found the values of both x and y, the algebraic process having been made possible by the fact that we had two equations connecting the unknown quantities.

The simultaneous equations above given might have been solved likewise by simply adding both equations together, thus:

Adding $\qquad\qquad x + y = 10$

and $\qquad\qquad\qquad x - y - 4,$

we have $\qquad\quad x + y + x - y = 14.$

Here $+y$ and $-y$ will cancel out, leaving

$$2x = 14,$$

$$x = 7.$$

Both of these processes are called *elimination*, the principal object in solving simultaneous equations being the elimination of unknown quantities until some equation is obtained in which only one unknown quantity appears.

We have seen that by simply adding two equations

we have eliminated one of the unknowns. But suppose the equations are of this type:

(1) $3x + 2y = 12$,
(2) $x + y = 5$.

Now we can proceed to solve these equations in one of two ways: first, to find the value of x in each equation and then equate these values of x, thus obtaining an equation where only y appears as an unknown quantity. But suppose we are trying to eliminate x from these equations by addition; it will be seen that adding will not eliminate x, nor even will subtraction eliminate it. If, however, we multiply equation (2) by 3, it becomes

$$3x + 3y = 15.$$

Now, when this is subtracted from equation (1), thus ·

$$\begin{array}{r} 3x + 2y = 12 \\ \underline{3x + 3y = 15} \\ -y = -3 \end{array}$$

the terms in x, $+3x$ and $+3x$ respectively, will eliminate, $3y$ minus $2y$ leaves $-y$, and $12 - 15$ leaves -3,

or $\qquad -y = -3$,

therefore $\qquad +y = +3$.

Just as in order to find the value of two unknowns two distinct and separate equations are necessary to express relations between these unknowns, likewise to find the value of the unknowns in equations containing three unknown quantities, three distinct and separate equations are necessary. Thus, we may have the equations

$$(1) \quad x + y + z = 6,$$
$$(2) \quad x - y + 2z = 1,$$
$$(3) \quad x + y - z = 4.$$

We now combine any two of these equations, for instance the first and the second, with the idea of eliminating one of the unknown quantities, as x. Subtracting equation (2) from (1), we will have

$$(4) \quad 2y - z = 5.$$

Now taking any other two of the equations, such as the second and the third, and subtracting one from the other, with a view to eliminating x, and we have

$$(5) \quad -2y + 3z = -3.$$

We now have two equations containing two unknowns, which we solve as before explained. For instance, adding them, we have

$$2z = 2,$$
$$z = 1.$$

Substituting this value of z in equation (4), we have

$$2y - 1 = 5,$$
$$2y = 6,$$
$$y = 3.$$

Substituting both of these values of z and y in equation (1), we have

$$x + 3 + 1 = 6,$$

$$x = 2.$$

Thus we see that with three unknowns three distinct and separate equations connecting them are necessary in order that their values may be found. Likewise with four unknowns four distinct and separate equations showing relations between them are necessary. In each case where we have a larger number than two equations, we combine the equations together two at a time, each time eliminating one of the unknown quantities, and, using the resultant equations, continue in the same course until we have finally resolved into one final equation containing only one unknown. To find the value of the other unknowns we then work backward, substituting the value of the one unknown found in an equation containing two unknowns, and both of these in an equation containing three unknowns, and so on.

The solution of simultaneous equations is very important, and the student should practice on this subject until he is thoroughly familiar with every one of these steps.

Solve the following problems:

1.
$$2x + y = 8,$$
$$2y - x = 6.$$

2. $$x + y = 7,$$
$$3x - y = 13.$$

3. $$4x = y + 2,$$
$$x + y = 3.$$

4. Find the value of x, y and z in the following equations:

$$x + y + z = 10,$$
$$2x + y - z = 9,$$
$$x + 2y + z = 12.$$

5. Find the value of x, y and z in the following equations:

$$2x + 3y + 2z = 20,$$
$$x + 3y + z = 13,$$
$$x + y + 2z = 13.$$

6. $$\frac{x}{3} + y = 10,$$
$$y + \frac{x}{5} = y - 3.$$

7. $$\frac{x}{4} + \frac{y}{3a} = 100x + a \text{ if } a = 8,$$
$$\frac{2x}{5} = y + 10.$$

8. $$3x + y = 15,$$
$$x = 6 + 7y.$$

9. $$\left. \begin{array}{l} \dfrac{9x}{a + b} = \dfrac{y}{a - b} - 7 \\[2mm] x + y = 5 \end{array} \right\} \text{ if } \begin{array}{l} a = 6, \\ b = 5. \end{array}$$

10. $$3x - y + 6x = 8,$$
$$y - 10 + 4y = x.$$

CHAPTER VIII

FUNDAMENTALS OF ALGEBRA

Quadratic Equations

THUS far we have handled equations where the unknown whose value we were solving for entered the equation in the first power. Suppose, however, that the unknown entered the equation in the second power; for instance, the unknown x enters the equation thus,

$$x^2 = 12 - 2x^2$$

In solving this equation in the usual manner we obtain

$$3x^2 = 12,$$
$$x^2 = 4.$$

Taking the square root of both sides,

$$x = \pm 2.$$

We first obtained the value of x^2 and then took the square root of this to find the value of x. The solution of such an equation is seen to be just as simple in every respect as a simple equation where the unknown did not appear as a square. But suppose that we have such an equation as this:

$$4x^2 + 8x = 12.$$

We see that none of the processes thus far discussed will do. We must therefore find some way of grouping

x^2 and x together which will give us a single term in x when we take the square root of both sides; this device is called " Completing the square in x."

It consists as follows: Group together all terms in x^2 into a single term, likewise all terms containing x into another single term. Place these on the left-hand side of the equation and everything else on the right-hand side of the equation. Now divide through by the coefficient of x^2. In the above equation this is 4. Having done this, add to the right-hand side of the equation the square of one-half of the coefficient of x. If this is added to one side of the equation it must likewise be added to the other side of the equation. Thus:

$$4x^2 + 8x = 12.$$

Dividing through by the coefficient of x^2, namely 4, we have

$$x^2 + 2x = 3.$$

Adding to both sides the square of one-half of the coefficient of x, which is 2 in the term $2x$,

$$x^2 + 2x + 1 = 3 + 1.$$

The left-hand side of this equation has now been made into the perfect square of $x + 1$, and therefore may be expressed thus:

$$(x + 1)^2 = 4.$$

Now taking the square root of both sides we have

$$x + 1 = \pm 2.$$

Therefore, using the plus sign of 2, we have

$$x = 1.$$

Using the minus sign of 2 we have

$$x = -3.$$

The student will note that there must, in the nature of the case, be two distinct and separate roots to a quadratic equation, due to the plus and minus signs above mentioned.

To recapitulate the preceding steps, we have:

(1) Group all the terms in x^2 and x on one side of the equation alone, placing those in x^2 first.

(2) Divide through by the coefficient of x^2.

(3) Add to both sides of the equation the square of one-half of the coefficient of the x term.

(4) Take the square root of both sides (the left-hand side being a perfect square). Then solve as for a simple equation in x.

Example: Solve for x in the following equation:

$$4x^2 = 56 - 20x,$$
$$4x^2 + 20x = 56,$$
$$x^2 + 5x = 14,$$
$$x^2 + 5x + \frac{25}{4} = 14 + \frac{25}{4},$$
$$x^2 + 5x + \frac{25}{4} = \frac{81}{4},$$
$$\left(x + \frac{5}{2}\right)^2 = \frac{81}{4}.$$

Taking the square root of both sides we have

$$x + \frac{5}{2} = \pm \frac{9}{2}.$$

$$x = \pm \frac{9}{2} - \frac{5}{2}.$$

$$x = 2 \text{ or } -7.$$

Example: Solve for x in the following equation:

$$2x^2 - 4x + 5 = x^2 + 2x - 10 - 3x^2 + 33,$$

$$2x^2 - x^2 + 3x^2 - 4x - 2x = 33 - 10 - 5,$$

$$4x^2 - 6x = 18,$$

$$x^2 - \frac{6x}{4} = \frac{18}{4},$$

$$x^2 - \frac{3x}{2} = \frac{18}{4},$$

$$x^2 - \frac{3x}{2} + \frac{9}{16} = \frac{18}{4} + \frac{9}{16}$$

$$\left(x - \frac{3}{4}\right)^2 = \frac{72}{16} + \frac{9}{16},$$

$$\left(x - \frac{3}{4}\right)^2 = \frac{81}{16},$$

$$x - \frac{3}{4} = \pm \frac{9}{4},$$

$$x = \pm \frac{9}{4} + \frac{3}{4},$$

$$x = +3 \text{ or } -1\tfrac{1}{2}.$$

Solving an Equation which Contains a Root. — Frequently we meet with an equation which contains a square or a cube root. In such cases it is necessary to get rid of the square or cube root sign as quickly as possible. To do this the root is usually placed on one side of the equation by itself, and then both sides are squared or cubed, as the case may be, thus:

Example: Solve the equation

$$\sqrt{2x + 6} + 5a = 10.$$

Solving for the root, we have

$$\sqrt{2x + 6} = 10 - 5a.$$

Now squaring both sides we have

$$2x + 6 = 100 - 100a + 25a^2,$$

or,

$$2x = 25a^2 - 100a + 100 - 6,$$

$$x = \frac{25a^2 - 100a + 94}{2}.$$

In any event, our prime object is first to get the square-root sign on one side of the equation by itself if possible, so that it may be removed by squaring.

Or the equation may be of the type

$$2a + 1 = \frac{4}{\sqrt{a - x}}.$$

Squaring both sides we have

$$4a^2 + 4a + 1 = \frac{16}{a - x}.$$

Clearing fractions we have

$$4\,a^2x - 4\,ax - x + 4\,a^3 + 4\,a^2 + a = 16,$$
$$x\,(4\,a^2 + 4\,a + 1) = -4\,a^3 - 4\,a^2 - a + 16,$$
$$x = \frac{4\,a^3 + 4\,a^2 + a - 16}{4\,a^2 + 4\,a + 1}.$$

PROBLEMS

Solve the following equations for the value of x:

1. $5\,x^2 - 15\,x = -10.$

2. $3\,x^2 + 4\,x + 20 = 44.$

3. $2\,x^2 + 11 = x^2 + 4\,x + 7.$

4. $x^2 + 4\,x = 2\,x + 2\,x^2 - 8.$

5. $7\,x + 15 - x^2 = 3\,x + 18.$

6. $x^4 + 2\,x^2 = 24.$

7. $x^2 + \dfrac{5\,x}{a} + 6\,x^2 = 10.$

8. $\dfrac{x^2}{a} + \dfrac{x}{b} - 3 = 0.$

9. $14 + 6\,x = 4\dfrac{x^2}{2} + \dfrac{2\,x}{a} \cdot 7.$

10. $\dfrac{x^2}{a+b} - 3\,x = 2.$

11. $3\,x^2 + 5\,x - 15 = 0.$

12. $(x + 2)^2 + 2\,(x + 2) = -1.$

13. $(x - 3)^2 - 10\,x + 7 = 0.$

14. $(x - a)^2 - (x + a)^2 = 3.$

15. $\dfrac{x+a}{x-a} + \dfrac{x+b}{x-b} = 2.$

16. $\dfrac{3x+7}{2} - \dfrac{x+2}{6} = \dfrac{12}{x+}$.

17. $\dfrac{x^2-2}{4x} = \dfrac{x+3+2x}{8}$.

18. $\dfrac{x^2-x-1}{4} = x^2+6$.

19. $8 = \dfrac{64}{\sqrt{x+1}}$.

20. $\sqrt{x+a} + 10a = 15$.

21. $\dfrac{x}{a} = \sqrt{x+1}$.

22. $3x+5 = 2 + \sqrt{3x+4}$.

CHAPTER IX

VARIATION

THIS is a subject of the utmost importance in the mathematical education of the student of science. It is one to which, unfortunately, too little attention is paid in the average mathematical textbook. Indeed, it is not infrequent to find a student with an excellent mathematical training who has but vaguely grasped the notions of variation, and still it is upon variation that we depend for nearly every physical law.

Fundamentally, variation means nothing more than finding the constants which connect two mutually varying quantities. Let us, for instance, take wheat and money. We know in a general way that the more money we have the more wheat we can purchase. This is a *variation* between wheat and money. But we can go no further in determining exactly how many bushels of wheat a certain amount of money will buy before we establish some definite constant relation between wheat and money, namely, the price per bushel of wheat. This price is called the *Constant* of the variation. Likewise, whenever two quantities are varying together, the movement of one depending absolutely upon the movement of the other, it is impossible to find out exactly what value of one corresponds with a given value of the other at any time, unless we know exactly what constant relation subsists between the two.

Where one quantity, a, *varies as* another quantity, namely, increases or decreases in value as another quantity, b, we represent the fact in this manner:

$$a \propto b.$$

Now, wherever we have such a relation we can immediately write

$$a = \text{some constant} \times b,$$

$$a = K \times b.$$

If we observe closely two corresponding values of a and b, we can substitute them in this equation and find out the value of this constant. This is the process which the experimenter in a laboratory has resorted to in deducing all the laws of science.

Experimentation in a laboratory will enable us to determine, not one, but a long series of corresponding values of two varying quantities. This series of values will give us an idea of the nature of their variation. We may then write down the variation as above shown, and solve for the *constant*. This *constant* establishes the relation between a and b at all times, and is therefore all-important. Thus, suppose the experimenter in a laboratory observes that by suspending a weight of

100 pounds on a wire of a certain length and size it stretched one-tenth of an inch. On suspending 200 pounds he observes that it stretches two-tenths of an inch. On suspending 300 pounds he observes that it stretches three-tenths of an inch, and so on. He at once sees that there is a constant relation between the elongation and the weight producing it. He then writes ·

Elongation $\propto$ weight.

Elongation = some constant $\times$ weight.

$E = K \times W$.

Now this is an equation. Suppose we substitute one of the sets of values of elongation and weight, namely,

3 of an inch and 300 lbs.

We have $\qquad$ $3 = K \times 300$.

Therefore $\qquad$ $K = .001$.

Now, this is an absolute constant for the stretch of that wire, and if at any time we wish to know how much a certain weight, say 500 lbs., will stretch that wire, we simply have to write down the equation

$$E = K \times W.$$

Substituting $\qquad$ elong. $= .001 \times 500$,

and we have $\qquad$ elong. $= .5$ of an inch.

Thus, in general, the student will remember that where two quantities vary as each other we can change this *variation*, which cannot be handled mathematically, into an *equation* which can be handled with absolute definiteness and precision by simply inserting a constant into the variation.

Inverse Variation.— Sometimes we have one quantity increasing at the same rate that another decreases; thus, the pressure on a certain amount of air increases as its volume is decreased, and we write

$$v \propto \frac{1}{p},$$

then
$$v = K \times \frac{1}{p}.$$

Wherever one quantity increases as another decreases, we call this an *inverse variation*, and we express it in the manner above shown. Frequently one quantity varies as the square or the cube or the fourth power of the other ; for instance, the area of a square varies as the square of its side, and we write

$$A \propto b^2,$$

or,
$$A = Kb^2.$$

Again, one quantity may vary inversely as the square of the other, as, for example, the intensity of light, which varies inversely as the square of the distance from its source, thus:

$$A \propto \frac{1}{d^2},$$

or,
$$A = K\frac{1}{d^2}.$$

Grouping of Variations. — Sometimes we have a quantity varying as one quantity and also varying as another quantity. In such cases we may group these two variations into a single variation. Thus, we say that

$$a \propto b,$$

also $\qquad a \propto c,$

then $\qquad a \propto b \times c,$

or, $\qquad a = K \times b \times c.$

This is obviously correct; for, suppose we say that the weight which a beam will sustain in end-on compression varies directly as its width, also directly as its depth, we see at a glance that the weight will vary as the cross-sectional area, which is the product of the *width* by the *depth*.

Sometimes we have such variations as this:

$$a \propto b,$$

also $\qquad a \propto \dfrac{1}{c},$

then $\qquad a \propto \dfrac{b}{c}.$

This is practically the same as the previous case, with the exception that instead of two direct variations we have one direct and one inverse variation.

There is much interesting theory in variation, which, however, is unimportant for our purposes and which

I will therefore omit. If the student thoroughly masters the principles above mentioned he will find them of inestimable value in comprehending the deduction of scientific equations.

PROBLEMS

1. If $a \propto b$ and we have a set of values showing that when $a = 500$, $b = 10$, what is the constant of this variation?

2. If $a \propto b^2$, and the constant of the variation is 2205, what is the value of b when $a = 5$?

3. $a \propto b$; also $a \propto \dfrac{1}{c}$. or, $a \propto \dfrac{b}{c}$ If we find that when $a = 100$, then $b = 5$ and $c = 3$, what is the constant of this variation?

4. $a \propto bc$. The constant of the variation equals 12. What is the value of a when $b = 2$ and $c = 8$?

5. $a = K \times \dfrac{b}{c}$ If $K = 15$ and $a = 6$ and $b = 2$, what is the value of c?

CHAPTER X

SOME ELEMENTS OF GEOMETRY

In this chapter I will attempt to explain briefly some elementary notions of geometry which will materially aid the student to a thorough understanding of many physical theories. At the start let us accept the following axioms and definitions of terms which we will employ.

Axioms and Definitions :

I. Geometry is the science of space.

II. There are only three fundamental directions or dimensions in space, namely, *length, breadth* and *depth.*

III. A geometrical *point* has theoretically no dimensions.

IV. A geometrical *line* has theoretically only one dimension, — *length.*

V. A geometrical *surface* or *plane* has theoretically only two dimensions, namely, *length* and *breadth.*

VI. A geometrical *body* occupies space and has three dimensions, — *length, breadth* and *depth.*

VII. An angle is the *opening* or *divergence* between two straight lines which cut or *intersect* each other; thus, in Fig. 1,

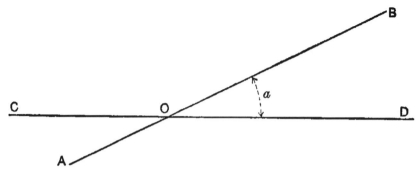

Fig 1.

∡ α is an angle between the lines AB and CD, and may be expressed thus, ∡ α or ∡ BOD.

VIII. When two lines lying in the same surface or plane are so drawn that they never approach or retreat from each other, no matter how long they are actually extended, they are said to be *parallel;* thus, in Fig. 2,

A   B

O   D

Fig. 2.

the lines AB and CD are parallel.

IX. A definite portion of a surface or plane bounded by lines is called a *polygon;* thus, Fig. 3 shows a polygon.

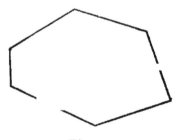

Fig. 3.

X. A *polygon* bounded by three sides is called a *triangle* (Fig. 4).

Fig. 4.

XI. A *polygon* bounded by four sides is called a *quadrangle* (Fig. 5), and if the opposite sides are parallel, a *parallelogram* (Fig. 6).

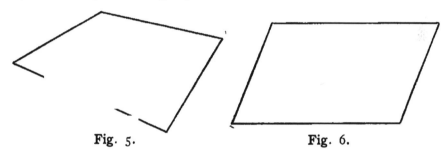

Fig. 5.   Fig. 6.

XII. When a line has revolved about a point until it has swept through a complete circle, or 360°, it comes back to its original position. When it has revolved one quarter of a circle, or 90°, away from its original position, it is said to be at *right angles* or *perpendicular* to its original

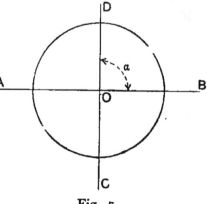

Fig. 7.

position; thus, the angle α (Fig. 7) is a *right angle*

between the lines AB and CD, which are perpendicular to each other.

XIII. An angle less than a right angle is called an *acute angle*.

XIV. An angle greater than a right angle is called an *obtuse angle*.

XV. The addition of two right angles makes a straight line.

XVI. Two angles which when placed side by side or added together make a right angle, or 90°, are said to be *complements* of each other; thus, ∠ 30° and ∠ 60° are *complementary angles*.

XVII. Two angles which when added together form 180°, or a straight line, are said to be *supplements* of each other; thus, ∠ 130° and ∠ 50° are *supplementary angles*.

XVIII. When one of the inside angles of a triangle is a right angle, it is called a *right-angle triangle* (Fig. 8),

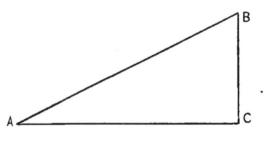

Fig. 8.

and the side AB opposite the right angle is called its *hypothenuse.*

XIX. A rectangle is a parallelogram whose angles are all right angles (Fig. 9a), and a square is a rectangle whose sides are all equal (**Fig. 9**).

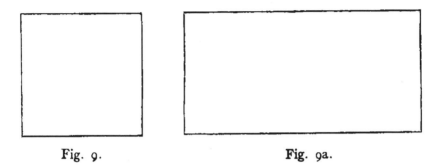

Fig. 9.   Fig. 9a.

XX. A circle is a curved line, all points of which are *equally distant* or *equidistant* from a fixed point called a center (**Fig. 10**).

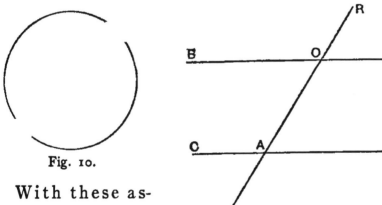

Fig. 10.

Fig. 11.

With these assumptions we may now proceed. Let us look at Fig. 11. *BM* and *CN* are parallel lines cut by the *common transversal* or intersecting line *RS.* It is seen at a glance that the ⊁ *ROM* and ⊁ *BOA*, called *vertical angles*, are equal; likewise ⊁ *ROM* and

∡ *RAN*, called *exterior interior angles*, are equal; likewise ∡ *BOA* and ∡ *RAN*, called *opposite interior angles*, are equal. These facts are actually proved by placing one on the other, when they will coincide exactly. The ∡ *ROM* and ∡ *BOR* are supplementary, as their sum forms the straight line *BM*, or 180°. Likewise ∡ *ROM* and ∡ *MOS*, or ∡ *NAS*, are supplementary.

In general, we have this rule: When the *corresponding sides of any two angles are parallel to each other, the angles are either equal or supplementary.*

Triangles. — Let us now investigate some of the properties of the triangle *ABC* (Fig. 12). Through *A*

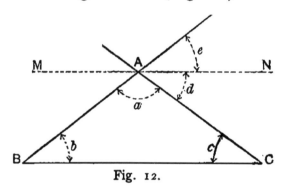

Fig. 12.

draw a line, *MN*, parallel to *BC*. At a glance we see that the sum of the angles *a*, *d*, and *e* is equal to 180°, or two right angles, —

$$∡\, a + ∡\, d + ∡\, e = 180°$$

But ∡ *c* is equal to ∡ *d*, and ∡ *b* is equal to ∡ *e*, as previously seen; therefore we have

$$∡\, a + ∡\, c + ∡\, b = 180°.$$

This demonstration proves the fact that the sum of all the inside or interior angles of any triangle is equal to 180°, or, what is the same thing, two right angles. Now, if the triangle is a right triangle and one of its angles is itself a right angle, then the *sum of the two remaining angles must be equal to one right angle, or* 90°. This fact should be most carefully noted, as it is very important.

When we have two triangles with all the angles of the one equal to the corresponding angles of the other, as in Fig. 13, they are called *similar triangles.*

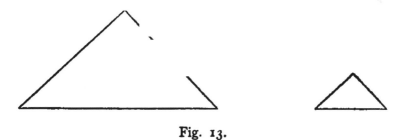

Fig. 13.

When we have two triangles with all three sides of the one equal to the corresponding sides of the other, they are equal to each other (Fig. 14), for they may be

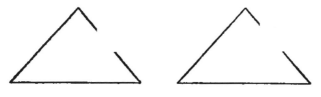

Fig. 14.

perfectly superposed on each other. In fact, the two triangles are seen to be equal if two sides and the included angle of the one are equal to two sides and the included angle of the other; or, if one side and two angles of the one are equal to one side and the corresponding angles respectively of the other; or, if one side and the angle opposite to it of the one are equal to one side and the corresponding angle of the other.

Projections. — The projection of any given tract, such as *AB* (Fig. 15), upon a line, such as *MN*, is that space,

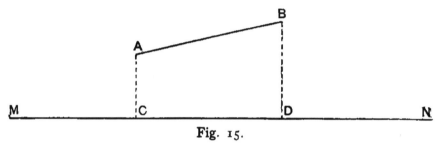

Fig. 15.

CD, on the line *MN* bounded by two lines drawn from *A* and *B* respectively perpendicular to *MN*.

Rectangles and Parallelograms. — A line drawn between opposite corners of a parallelogram is called a diagonal; thus, *AC* is a diagonal in Fig. 16. It is along

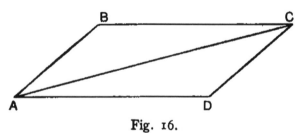

Fig. 16.

this diagonal that a body would move if pulled in the direction of *AB* by one force, and in the direction *AD* by another, the two forces having the same relative magnitudes as the relative lengths of *AB* and *AD*. This fact is only mentioned here as illustrative of one of the principles of mechanics.

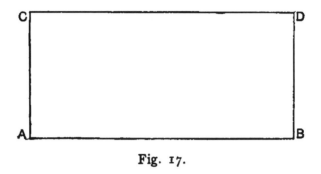

Fig. 17.

The area of a rectangle is equal to the product of the length by the breadth; thus, in Fig. 17,

$$\text{Area of } ABDC - AB \times AC.$$

This fact is so patent as not to need explanation.

Suppose we have a parallelogram (Fig. 18), however, what is its area equal to ?

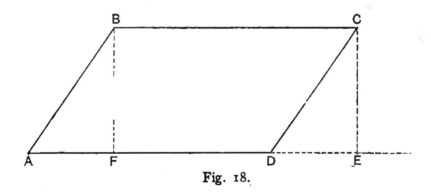

Fig. 18.

The perpendicular distance *BF* between the sides *BC* and *AD* of a parallelogram is called its *altitude*. Extend the base *AD* and draw *CE* perpendicular to it.

Now we have the rectangle $BCEF$, whose area we know to be equal to $BC \times BF$. But the triangles ABF and DCE are equal (having 2 sides and 2 angles mutually equal), and we observe that the rectangle is nothing else than the parallelogram with the triangle ABF chipped off and the triangle DCE added on, and since these are equal, the rectangle is equal to the parallelogram, which then has the same area as it; or,

Area of parallelogram $ABCD = BC \times BF$.

If, now, we consider the area of the triangle ABC (Fig. 19), we see that by drawing the lines AD and CD

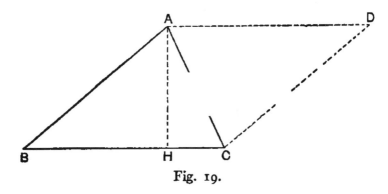

Fig. 19.

parallel to BC and AB respectively, we have the parallelogram $BADC$, and we observe that the triangles ABC and ADC are equal. Therefore triangle ABC equals one-half of the parallelogram, and since the area of this is equal to $BC \times AH$, then the

Area of the triangle $ABC = \frac{1}{2} BC \times AH$,

which means that the area of a triangle is equal to one-half of the product of the base by the altitude.

Circles. — Comparison between the lengths of the diameter and circumference of a circle (Fig. 20) made with the utmost care shows that the circumference is 3.1416 times as long as the diameter. This constant, 3.1416, is usually expressed by the Greek letter pi (π). Therefore, the circumference of a circle is equal to $\pi \times$ the diameter.

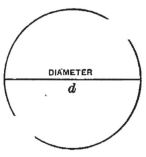

Fig. 20.

$$\text{circum.} = \pi d,$$

$$\text{circum.} = 2\,\pi r$$

if r, the radius, is used instead of the diameter.

The area of a segment of a circle (Fig. 21), like the area of a triangle, is equal to $\frac{1}{2}$ of the product of the base by the altitude, or $\frac{1}{2} a \times r$.

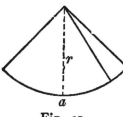

Fig. 21.

This comes from the fact that the segment may be divided up into a very large number of small segments whose bases, being very small, have very little curvature, and may therefore be considered as small triangles. Therefore, if we consider the whole circle, where the length of the arc is $2\,\pi r$, the area is

$$\frac{1}{2} \times 2\,\pi r \times r = \pi r^2,$$

$$\text{Area circle} = \pi r^2.$$

I will conclude this chapter by a discussion of one of
the most important properties of the right-angle triangle, namely, that the_square erected on its hypothenuse is equal to the sum of the squares erected on
its other two sides; that is, that in the triangle ABC
(Fig. 22) $\overline{AC}^2 = \overline{AB}^2 + \overline{BC}^2$

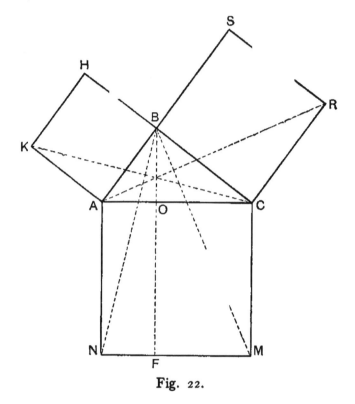

Fig. 22.

To prove $ANMC = BCRS + ABHK$,

or   length $\overline{AC}^2 -$ length $\overline{BC}^2 +$ length $\overline{AB}^2$.

This is a difficult problem and one of the most interesting and historic ones that the whole realm of mathematics can offer, therefore I will only suggest its solu-

tion and leave a little reasoning for the student himself to do.

$$\text{triangle } ARC = \text{triangle } BMC,$$
$$\text{triangle } ARC = \tfrac{1}{2} CR \times BC$$
$$= \tfrac{1}{2} \text{ of the square } BCRS,$$
$$\text{triangle } BCM = \tfrac{1}{2} CM \times CO$$
$$- \tfrac{1}{2} \text{ of rectangle } COFM.$$

Therefore

$$\tfrac{1}{2} \text{ of square } BCRS = \tfrac{1}{2} \text{ of rectangle } COFM,$$

or $$BCRS = COFM.$$

Similarly for the other side

$$ABHK = AOFN.$$

But

$$COFM + AOFN = \text{whole square } ACMN.$$

Therefore $$ACMN = BCRS + ABHK.$$

$$\overline{AC}^2 - \overline{BC}^2 + \overline{AB}^2.$$

PROBLEMS

1. What is the area of a rectangle 8 ft. long by 12 ft. wide ?

2. What is the area of a triangle whose base is 20 ft. and whose altitude is 18 ft.?

3. What is the area of a circle whose radius is 9 ft.?

4. What is the length of the hypothenuse of a right-angle triangle if the other two sides are respectively 6 ft. and 9 ft.?

5. What is the circumference of a circle whose diameter is 20 ft.?

6. The hypothenuse of a right-angle triangle is 25 ft. and one side is 18 ft.; what is the other side?

7. If the area of a circle is 600 sq. ft., what is its diameter?

8. The circumference of the earth is 25,000 miles; what is its diameter in miles?

9. The area of a triangle is 30 sq. ft. and its base is 8 ft.; what is its altitude?

10. The area of a parallelogram is 100 sq. feet and its base is 25 ft.; what is its altitude?

CHAPTER XI

ELEMENTARY PRINCIPLES OF TRIGONOMETRY

TRIGONOMETRY is the science of angles; its province is to teach us how to measure and employ angles with the same ease that we handle lengths and areas.

In a previous chapter we have defined an angle as the opening or the divergence between two intersecting lines, AB and CD (Fig. 23). The next question is, How

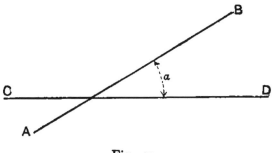

Fig. 23.

are we going to measure this angle? We have already seen that we can do this in one way by employing degrees, a complete circle being 360°. But there are many instances which the student will meet later on where the use of degrees would be meaningless. It is then that certain constants connected with the angle, called its *functions*, must be resorted to. Suppose we have the angle α shown in Fig. 24. Now let us choose

75

a point anywhere either on the line AB or CD ; for instance, the point P. From P drop a line which will

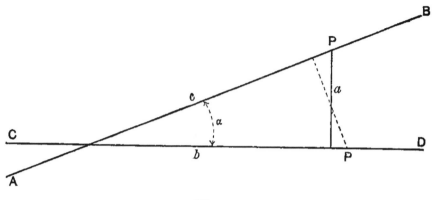

Fig. 24.

be perpendicular to CD. This gives us a right-angle triangle whose sides we may call a, b and c respectively. We may now define the following *functions* of the ∡ α ·

$$\text{sine } \alpha = \frac{a}{c},$$

$$\text{cosine } \alpha - \frac{b}{c},$$

$$\text{tangent } \alpha = \frac{a}{b},$$

which means that the *sine* of an angle is obtained by dividing the side opposite to it by the hypothenuse; the *cosine*, by dividing the side adjacent to it by the hypothenuse; and the *tangent*, by dividing the side opposite by the side adjacent.

These values, sine, cosine and tangent, are therefore nothing but ratios, — pure numbers, — and under no circumstances should be taken for anything else.   This is one of the greatest faults that I have to find with many texts and handbooks in not insisting on this point.

Looking at Fig. 24, it is evident that no matter where I choose P, the values of the sine, cosine and tangent will be the same; for if I choose P farther out on the line I will increase c, but at the same time a will increase in the same proportion, the quotient of $\dfrac{a}{c}$ being always the same wherever P may be chosen.

Likewise $\dfrac{b}{c}$ and $\dfrac{a}{b}$ will always remain constant.   The *sine, cosine* and *tangent* are therefore always *fixed* and *constant* quantities for any given angle.   I might have remarked that if P had been chosen on the line CD and the perpendicular drawn to AB, as shown by the dotted lines (Fig. 24), the hypothenuse and adjacent side simply exchange places, but the value of the sine, cosine and tangent would remain the same.

Since these functions, namely, sine, cosine and tangent, of any angle remain the same at all times, they become very convenient handles for employing the angle.   The sines, cosines and tangents of all angles of every size may be actually measured and computed with great care once and for all time, and then arranged in tabulated form, so that by referring to this table one can immediately find the sine, cosine or tangent of any angle; or, on the other hand, if a certain value said to be the sine, cosine or tangent of an unknown angle is given, the angle that it corresponds to may be found from the table. Such a table may be found at the end of this book, giving the sines, cosines and tangents of all angles taken 6 minutes apart. Some special compilations of these tables give the values for all angles taken only one minute apart, and some even closer, say 10 seconds apart.

On reference to the table, the sine of 10° is .1736, the cosine of 10° is .9848, the sine of 24° 36' is .4163, the cosine of 24° 36' is .9092. In the table of sines and cosines the decimal point is understood to be before every value, for, if we refer back to our definition of sine and cosine, we will see that these values can never be greater than 1; in fact, they will always be less than 1, since the hypothenuse c is always the longest side of the right angle and therefore a and b are always less than it. Obviously, $\frac{a}{c}$ and $\frac{b}{c}$, the values respectively of sine and cosine, being a smaller quantity divided by a larger, can never be greater than 1. Not so with the tangent; for angles between 0° and 45°, a is less than b, therefore $\frac{a}{b}$ is less than 1; but for angles between 45° and 90°, a is greater than b, and therefore $\frac{a}{b}$ is greater than 1. Thus, on reference to the table the tangent of 10° 24' is seen to be .1835, the tangent of 45° is 1, the tangent of 60° 30' is 1.7675.

Now let us work backwards. Suppose we are given .3437 as the sine of a certain angle, to find the angle. On reference to the table we find that this is the sine of 20° 6′, therefore this is the angle sought. Again, suppose we have .8878 as the cosine of an angle, to find the angle. On reference to the table we find that this is the angle 27° 24′. Likewise suppose we are given 3.5339 as the tangent of an angle, to find the angle. The tables show that this is the angle 74° 12′.

When an angle or value which is sought cannot be found in the tables, we must prorate between the next higher and lower values. This process is called *interpolation*, and is merely a question of proportion. It will be explained in detail in the chapter on Logarithms.

Relation of Sine and Cosine. — On reference to Fig. 25 we see that the sine $\alpha - \dfrac{a}{c}$, but if we take β, the other acute angle of the right-angle triangle, we see that cosine $\beta = \dfrac{a}{c}$.

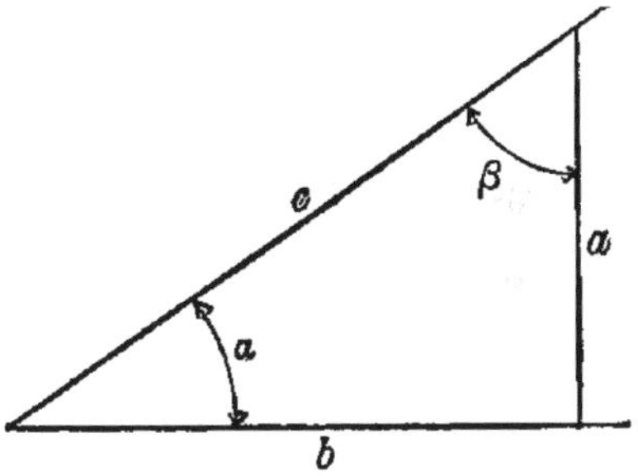

Fig. 25.

Remembering always the fundamental definition of sine and cosine, namely,

$$\text{sine} = \frac{\text{Opposite side}}{\text{Hypothenuse}},$$

$$\text{cosine} = \frac{\text{Adjacent side}}{\text{Hypothenuse}},$$

we see that the *cosine* β is equal to the same thing as the *sine* α, therefore

$$\text{sine } \alpha = \text{cosine } \beta.$$

Now, if we refer back to our geometry, we will remember that the sum of the three angles of a triangle − 180°, or two right angles, and therefore in a right-angle triangle $\angle \alpha + \angle \beta = 90°$, or 1 right angle. In other words $\angle \alpha$ and $\angle \beta$ are *complementary angles*. We then have the following general law: "*The sine of an angle is equal to the cosine of its complement*" Thus, if we have a table of *sines* or *cosines* from 0° to 90°, or *sines and cosines* between 0° and 45°, we make use of this principle. If we are asked to find the *sine* of 68° we may look for the *cosine* of (90° − 68°), or 22°; or, if we want the *cosine* of 68°, we may look for the *sine* of (90° − 68°), or 22°.

Other Functions. — There are some other functions of the angle which are seldom used, but which I will mention here, namely,

$$\text{Cotangent} - \frac{b}{a},$$

$$\text{Secant} = \frac{c}{b},$$

$$\text{Cosecant} - \frac{c}{a}$$

Other Relations of Sine and Cosine. — We have seen

that the sine $\alpha = \dfrac{a}{c}$ and the cosine $\alpha = \dfrac{b}{c}$ Also from geometry

$$a^2 + b^2 = c^2. \tag{1}$$

Dividing equation (1) by c^2 we have

$$\frac{a^2}{c^2} + \frac{b^2}{c^2} = 1.$$

But this is nothing but the square of the sine plus the square of the cosine of $\measuredangle\ \alpha$, therefore

$$(\text{sine } \alpha)^2 + (\text{cosine } \alpha)^2 = 1.$$

Other relations whose proof is too intricate to enter into now are

$$\text{sine } 2\alpha = 2 \sin \alpha \cos \alpha,$$
$$\cos 2\alpha = 1 - 2\sin^2 \alpha,$$

or
$$\cos 2\alpha = \cos^2 \alpha - \sin^2 \alpha.$$

Use of Trigonometry. — Trigonometry is invaluable in triangulation of all kinds. When two sides or one side and an acute angle of a right-angle triangle are

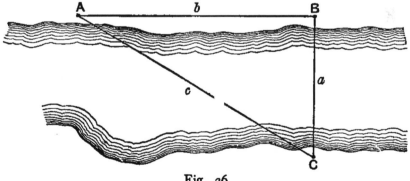

Fig. 26.

given, the other two sides can be easily found. Suppose we wish to measure the distance BC across the river in Fig. 26; we proceed as follows: First we lay off and measure the distance AB along the shore; then by means of a transit we sight perpendicularly across the river and erect a flag at C; then we sight from A to B and from A to C and determine the angle α. Now, as before seen, we know that

$$\text{tangent } \alpha = \frac{a}{b}.$$

Suppose b had been 1000 ft. and $\angle \alpha$ was 40°, then

$$\text{tangent } 40° = \frac{a}{1000}.$$

The tables show that the tangent of 40° is .8391;

then

$$.8391 = \frac{a}{1000},$$

therefore

$$a = 839.1 \text{ ft.}$$

Thus we have found the distance across the river to be 839.1 ft.

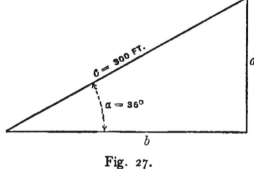

Fig. 27.

Likewise in Fig. 27, suppose $c = 300$ and $\angle \alpha = 36°$, to find a and b. We have

$$\text{sine } \alpha = \frac{a}{c},$$

or

$$\text{sine } 36° = \frac{a}{300}.$$

From the tables sine $36° = .5878$.

$$.5878 = \frac{a}{300},$$

$$a = .5878 \times 300,$$

or $\qquad\qquad a = 176.34 \text{ ft.}$

Likewise $\qquad$ cosine $\alpha = \dfrac{b}{c}$

From table, $\quad$ cosine $36' = .8090$,

therefore $\qquad\qquad .8090 = \dfrac{b}{300},$

or $\qquad\qquad b = 242.7 \text{ ft.}$

Now, if we had been told that $a = 225$ and $b = 100$ to find $\measuredangle \alpha$ and c, we would have proceeded thus:

$$\text{tangent } \alpha = \frac{a}{b}.$$

Therefore $\qquad$ tangent $\alpha = \dfrac{225}{100},$

$$\text{tangent } \alpha = 2.25 \text{ ft.}$$

The tables show that this corresponds to the angle $66° 4'$

Therefore $\qquad\qquad \alpha = 66° 4'.$

Now to find c we have

$$\sin \alpha = \frac{a}{c},$$

$$\sin 66° 4' = \frac{225}{c}.$$

From tables,   sine $66° 4' = .9140$,

therefore $.9140 = \dfrac{225}{c}$,

or $\underline{\quad} = \dfrac{225}{.9140} = 248.5$ ft.

And thus we may proceed, the use of a little judgment being all that is necessary to the solution of the most difficult problems of triangulation.

PROBLEMS

1. Find the sine, cosine and tangent of $32° 20'$.
2. Find the sine, cosine and tangent of $81° 24'$
3. What angle is it whose sine is .4320 ?
4. What angle is it whose cosine is .1836 ?
5. What angle is it whose tangent is .753 ?
6. What angle is it whose cosine is .8755 ?

In a right-angle triangle —

7. If $a = 300$ ft. and $\not\!\!\angle\, \alpha = 30°$, what are c and b ?
8. If $a = 500$ ft. and $b = 315$ ft., what are $\not\!\!\angle\, \alpha$ and c?
9. If $c = 1250$ ft. and $\not\!\!\angle\, \alpha = 80°$, what are b and a ?
10. If $b = 250$ ft. and $c = 530$ ft., what are $\not\!\!\angle\, \alpha$ and a ?

CHAPTER XII

LOGARITHMS

I HAVE inserted this chapter on logarithms because I consider a knowledge of them very essential to the education of any engineer.

Definition. — A logarithm is the power to which we must raise a given base to produce a given number. Thus, suppose we choose 10 as our base, we will say that 2 is the logarithm of 100, because we must raise 10 to the second power — in other words, square it — in order to produce 100. Likewise 3 is the logarithm of 1000, for we have to raise 10 to the third power (thus, 10^3) to produce 1000. The logarithm of 10,000 would then be 4, and the logarithm of 100,000 would be 5, and so on.

The *base* of the universally used *Common System* of logarithms is 10; of the *Naperian* or *Natural System*, ϵ or 2.7. The latter is seldom used.

We see that the logarithms of such numbers as 100, 1000, 10,000, etc., are easily detected; but suppose we have a number such as 300, then the difficulty of finding its logarithm is apparent. We have seen that 10^2 is 100, and 10^3 equals 1000, therefore the number 300, which lies between 100 and 1000, must have a logarithm which lies between the logarithms of 100 and 1000,

namely 2 and 3 respectively. Reference to a table of
logarithms at the end of this book, which we will explain later, shows that the logarithm of 300 is 2.4771,
which means that 10 raised to the 2.4771ths power will
give 300. The *whole number* in a logarithm, for example the 2 in the above case, is called the *characteristic;*
the decimal part of the logarithm, namely, .4771, is
called the *mantissa.* It will be hard for the student to
understand at first what is meant by raising 10 to a
fractional part of a power, but he should not worry
about this at the present time; as he studies more
deeply into mathematics the notion will dawn on him
more clearly.

We now see that every number has a logarithm, no
matter how large or how small it may be; every number
can be produced by raising 10 to some power, and this
power is what we call the *logarithm* of the number.
Mathematicians have carefully worked out and tabulated the logarithm of every number, and by reference
to these tables we can find the logarithm corresponding
to any number, or vice versa. A short table of logarithms is shown at the end of this book.

Now take the number 351.1400; we find its logarithm
is 2.545,479. Like all numbers which lie between 100
and 1000 its characteristic is 2. The numbers which
lie between 1000 and 10.000 have 3 as a characteristic;
between 10 and 100, 1 as a characteristic. We there-

fore have the rule that *the characteristic is always one less than the number of places to the left of the decimal point.* Thus, if we have the number 31875.12, we immediately see that the characteristic of its logarithm will be 4, because there are five places to the left of the decimal point. Since it is so easy to detect the characteristic, it is never put in logarithmic tables, the *mantissa* or *decimal* part being the only part that the tables need include.

If one looked in a table for a logarithm of 125.60, he would only find .09,899. This is only the *mantissa* of the logarithm, and he would himself have to insert the characteristic, which, being one less than the number of places to the left of the decimal point, would in this case be 2 ; therefore the logarithm of 125.6 is 2.09,899.

Furthermore, the *mantissæ* of the logarithms of 3.4546, 34.546, 345.46, 3454.6, etc., are all exactly the same. The characteristic of the logarithm is the only thing which the decimal point changes, thus:

$$\log 3.4546 = 0.538,398,$$
$$\log 34.546 = 1.538,398,$$
$$\log 345.46 = 2.538,398,$$
$$\log 3454.6 = 3.538,398,$$
$$\text{etc.}$$

Therefore, in looking for the logarithm of a number, first put down the *characteristic* on the basis of the above rules, then look for the *mantissa* in a table, neglecting the position of the decimal point altogether. Thus, if we are looking for the logarithm of .9840, we first write down the characteristic, which in this case would be −1 (there are no places to the left of the decimal point in this case, therefore one less than none is −1). Now look in a table of logarithms for the mantissa which corresponds to .9840, and we find this to be .993,083; therefore

$$\log .9840 = -1.993,083.$$

If the number had been 98.40 the logarithm would have been +1.993,083.

When we have such a number as .084, the characteristic of its logarithm would be −2, there being one less than no places at all to the left of its decimal point; for, even if the decimal point were moved to the right one place, you would still have no places to the left of the decimal point; therefore

$$\log .00,386 \ = - \ 3.586,587,$$
$$\log 38.6 \qquad - \ 1.586,587,$$
$$\log 386 \qquad - \ 2.586,587,$$
$$\log 386,000 = 5.586,587.$$

Interpolation. — Suppose we are asked to find the logarithm of 2468; immediately write down 3 as the characteristic. Now, on reference to the logarithmic

table at the end of this book, we see that the logarithms of 2460 and 2470 are given, but not 2468. Thus:

$$\log 2460 = 3.3909,$$
$$\log 2468 = ?$$
$$\log 2470 = 3.3927.$$

We find that the total difference between the two
given logarithms, namely 3909 and 3927, is 16, the
total difference between the numbers corresponding to
these logarithms is 10, the difference between 2460 and
2468 is 8; therefore the logarithm to be found lies $\frac{8}{10}$
of the distance across the bridge between the two given
logarithms 3909 and 3927. The whole distance across
is 16. $\frac{8}{10}$ of 16 is 12.8. Adding this to 3909 we have
3921.8; therefore

$$\log \text{ of } 2468 = 3.39,218.$$

Reference to column 8 in the interpolation columns to the
right of the table would have given this value at once.

Many elaborate tables of logarithms may be purchased
at small cost which make interpolation almost unneces-
sary for practical purposes.

Now let us work backwards and find the number if
we know its logarithm. Suppose we have given the
logarithm 3.6201. Referring to our table, we see that
the mantissa .6201 corresponds to the number 417; the
characteristic 3 tells us that there must be four places
to the left of the decimal point; therefore

3.6201 is the log of 4170.0.

Now, for interpolation we have the same principles aforesaid. Let us find the number whose log is −3.7304. In the table we find that

> log 7300 corresponds to the number 5370,
> log 7304 corresponds to the number   ?
> log 7308 corresponds to the number 5380.

Therefore it is evident that

> 7304 corresponds to 5375.

Now the characteristic of our logarithm is − 3; from this we know that there must be two zeros to the left of the decimal point; therefore

> 3.7304 is the log of the number .005375.

Likewise

> −2.7304 is the log of the number .05375,
> .7304 is the log of the number 5.375,
> 4.7304 is the log of the number 53,750.

Use of the Logarithm. — Having thoroughly understood the nature and meaning of a logarithm, let us investigate its use mathematically. It changes *multiplication* and *division* into *addition* and *subtraction; involution* and *evolution* into *multiplication* and *division.*

We have seen in algebra that

$$a^2 \times a^5 = a^{5+2}, \text{ or } a^7,$$

and that

$$\frac{a^8}{a^3} = a^{8-3}, \text{ or } a^5.$$

In other words, multiplication or division of like symbols was accomplished by adding or subtracting their exponents, as the case may be. Again, we have seen that

$$(a^2)^2 = a^4,$$

or
$$\sqrt[3]{a^6} = a^2.$$

In the first case a^2 squared gives a^4, and in the second case the cube root of a^6 is a^2; to raise a number to a power you multiply its exponent by that power; to find any root of it you divide its exponent by the exponent of the root. Now, then, suppose we multiply 336 by 5380; we find that

$$\log \text{ of } 336 = 10^{2.5263},$$

$$\log \text{ of } 5380 = 10^{3.7308}$$

Then 336×5380 is the same thing as $10^{2.5263} \times 10^{3.7308}$.

But $10^{2.5263} \times 10^{3.7308} - 10^{2.5263 + 3.7308} - 10^{6.2571}$

We have simply added the exponents, remembering that these exponents are nothing but the logarithms of 336 and 5380 respectively.

Well, now, what number is $10^{6.2571}$ equal to? Look ing in a table of logarithms we see that the mantissa .2571 corresponds to 1808; the characteristic 6 tells us that there must be seven places to the left of the decimal; therefore

$$10^{6.2571} = 1,808,000.$$

If the student notes carefully the foregoing he will see that in order to multiply 336 by 5380 we simply find their logarithms, add them together, getting another logarithm, and then find the number corresponding to this logarithm. Any numbers may be multiplied together in this simple manner; thus, if we multiply $217 \times 4876 \times 3.185 \times .0438 \times 890$, we have

$$
\begin{aligned}
\log 217 &= 2.3365 \\
\log 4876 &- 3.6880 \\
\log 3.185 &- .5031 \\
\log .0438 &= -2.6415^* \\
\log 890 &- 2.9494
\end{aligned}
$$

Adding we get   8.1185

We must now find the number corresponding to the logarithm 8.1185. Our tables show us that

8.1185 is the log of 131,380,000.

Therefore 131,380,000 is the result of the above multiplication.

To divide one number by another we subtract the logarithm of the latter from the logarithm of the former; thus, $3865 \div 735$:

$$
\begin{aligned}
\log 3865 &= 3.5872 \\
\log 735 &= 2.8663 \\
\hline
&.7209
\end{aligned}
$$

The tables show that .7209 is the logarithm of 5.259; therefore

$$3865 \div 735 = 5.259.$$

* The -2 does not carry its negativity to the mantissa.

As explained above, if we wish to square a number, we simply multiply its logarithm by 2 and then find what number the result is the logarithm of. If we had wished to raise it to the third, fourth or higher power, we would simply have multiplied by 3, 4 or higher power, as the case may be. Thus, suppose we wish to cube 9879; we have

$$\log 9897 = 3.9947$$
$$\underline{\qquad 3}$$
$$11.9841$$

11.9841 is the log of 964,000,000,000;

therefore 9879 cubed = 964,000,000,000.

Likewise, if we wish to find the square root, the cube root, or fourth root or any root of a number, we simply divide its logarithm by 2, 3, 4 or whatever the root may be; thus, suppose we wish to find the square root of 36,850, we have

$$\log 36{,}850 = 4.5664.$$
$$4.5664 \div 2 = 2.2832.$$

2.2832 is the log. of 191.98; therefore the square root of 36,850 is 191.98.

The student should go over this chapter very carefully, so as to become thoroughly familiar with the principles involved.

PROBLEMS

1. Find the logarithm of 3872.
2. Find the logarithm of 73.56.
3. Find the logarithm of .00988.
4. Find the logarithm of 41,267.
5. Find the number whose logarithm is 2.8236.
6. Find the number whose logarithm is 4.87175.
7. Find the number whose logarithm is −1.4385.
8. Find the number whose logarithm is −4.3821.
9. Find the number whose logarithm is 3.36175.
10. Multiply 2261 by 4335.
11. Multiply 6^{218} by 3998.
12. ′ Multiply 231.9 by 478.8 by 7613 by .921.
13. Multiply .00983 by .0291.
14. Multiply .222 by .00054.
15. Divide 27,683 by 856.
16. Divide 4337 by 38.88.
17. Divide .9286 by 28.75.
18. Divide .0428 by 1.136.
19. Divide 3995 by .003,337.
20. Find the square of 4291.
21. Raise 22.91 to the fourth power.
22. Raise .0236 to the third power.
23. Find the square root of 302,060.
24. Find the cube root of 77.85.
25. Find the square root of .087,64.
26. Find the fifth root of 226,170,000.

CHAPTER XIII

ELEMENTARY PRINCIPLES OF COÖRDINATE GEOMETRY

CoÖRDINATE Geometry may be called *graphic algebra,* or *equation drawing,* in that it depicts algebraic equations not by means of symbols and terms but by means of curves and lines. Nothing is more familiar to the engineer, or in fact to any one, than to see the results of machine tests or statistics and data of any kind shown graphically by means of curves. The same analogy exists between an algebraic equation and the curve which graphically represents it as between the verbal description of a landscape and its actual photograph; the photograph tells at a glance more than could be said in many thousands of words. Therefore the student will realize how important it is that he master the few fundamental principles of coördinate geometry which we will discuss briefly in this chapter.

An Equation. — When discussing equations we remember that where we have an equation which contains two unknown quantities, if we assign some numerical value to one of them we may immediately find the corresponding numerical value of the other; for example, take the equation

$$x = y + 4.$$

In this equation we have two unknown quantities, namely, x and y; we cannot find the value of either unless we know the value of the other. Let us say that $y = 1$; then we see that we would get a corresponding value, $x = 5$; for $y = 2$, $x = 6$; thus:

$$\text{If } y = 1, \text{ then } x = 5,$$
$$y = 2, \qquad x = 6,$$
$$y = 3, \qquad x = 7,$$
$$y = 4, \qquad x = 8,$$
$$y = 5, \qquad x = 9, \text{ etc.}$$

The equation then represents the relation in value existing between x and y, and for any specific value of x we can find the corresponding specific value of y. Instead of writing down, as above, a list of such corresponding values, we may show them graphically thus: Draw two lines perpendicular to each other; make one of them the x line and the other the y line. These two lines are called axes. Now draw parallel to these axes equi-spaced lines forming cross-sections, as shown in Fig. 28, and letter the intersections of these lines with the axes 1, 2, 3, 4, 5, 6, etc., as shown.

Now let us plot the corresponding values, $y = 1$, $x = 5$. This will be a point 1 *space* up on the y axis and 5 spaces out on the x axis, and is denoted by letter A in the figure. In plotting the corresponding values $y = 2$, $x = 6$, we get the point B; the next set of values

gives us the point C, the next D, and so on. Suppose
we draw a line through these points; this line, called
the curve of the equation, tells everything in a graphical

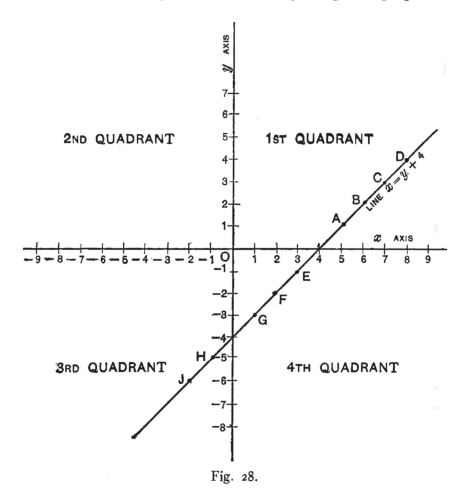

Fig. 28.

way that the equation does algebraically. If this line
has been drawn accurately we can from it find out at
a glance what value of y corresponds to any given value
of x, and vice versa. For example, suppose we wish to
see what value of y corresponds to the value $x = 6\frac{1}{2}$;

we run our eyes along the x axis until we come to $6\frac{1}{2}$, then up until we strike the curve, then back upon the y axis, where we note that $y = 2\frac{1}{2}$.

Negative Values of x and y. — When we started at o and counted 1, 2, 3, 4, etc., to the right along the x axis, we might just as well have counted to the left, -1, -2, -3, 4, etc. (Fig. 28), and likewise we might have counted downwards along the y axis, -1, 2, -3, -4, etc. The values, then, to the left of o on the x axis and below o on the y axis are the negative values of x and y. Still using the equation $x = y + 4$, let us give the following values to y and note the corresponding values of x in the equation $x = y + 4$:

$$\text{If } y = 0, \quad \text{then } x = 4,$$
$$y = -1, \quad x = 3,$$
$$y = -2, \quad x = 2,$$
$$y = -3, \quad x = 1,$$
$$y = -4, \quad x = 0,$$
$$y = -5, \quad x = -1,$$
$$y = -6, \quad x = -2,$$
$$y = -7, \quad x = -3.$$

The point $y = 0$, $x = 4$ is seen to be on the x axis at the point 4. The point $y = -1$, $x = 3$ is at point E, that is, 1 below the x axis and 3 to the right of the y axis. The points $y = -2$, $x = 2$ and $y = -3$, $x = 1$ are seen to be respectively points F and G. Point

$y = -4$, $x = 0$ is zero along the x axis, and is therefore at -4 on the y axis. Point $y = -5$, $x = -1$
is seen to be 5 below o on the y axis and 1 to
the left of o along the x axis (both x and y are now
negative), namely, at the point H. Point $y = -6$,
$x = -2$ is at J, and so on.

The student will note that all points in the first
quadrant have positive values for both x and y, all
points in the second quadrant have positive values for
y (being all above o so far as the y axis is concerned),
but negative values for x (being to the left of o), all
points in the third quadrant have negative values
for both x and y, while all points in the fourth quadrant
have positive values of x and negative values of y.

Coördinates. — The corresponding x and y values of
a point are called its *coördinates*, the *vertical* or y value
is called its *ordinate*, while the *horizontal* or x value is
called the *abscissa;* thus at point A, $x = 5$, $y = 1$, here
5 is called the *abscissa*, while 1 is called the *ordinate* of
point A. Likewise at point G, where $y = -3$, $x = 1$,
here -3 is the *ordinate* and 1 the *abscissa* of G.

Straight Lines. — The student has no doubt observed
that all points plotted in the equation $x = y + 4$ have
fallen on a straight line, and this will always be the case
where both of the unknowns (in this case x and y) enter
the equation only in the first power; but the line will
not be a straight one if either x or y or both of them

enter the equation as a square or as a higher power; thus, $x^2 = y + 4$ will not plot out a straight line because we have x^2 in the equation. Whenever both of the unknowns in the equation which we happen to be plotting (be they x and y, a and b, x and a, etc.) enter the equation in the first power, the equation is called a *linear equation*, and it will always plot a straight line; thus, $3x + 5y = 20$ is a linear equation, and if plotted will give a straight line.

Conic Sections. — If either or both of the unknown quantities enter into the equation in the second power, and no higher power, the equation will always represent one of the following curves: a *circle* or an *ellipse*, a *parabola* or an *hyperbola*. These curves are called the conic sections. A typical equation of a circle is $x^2 + y^2 = r^2$; a typical equation of a parabola is $y^2 = 4qx$; a typical equation of a hyperbola is $x^2 - y^2 = r^2$, or, also, $xy = c^2$.

It is noted in every one of these equations that we have the second power of x or y, except in the equation $xy = c^2$, one of the equations of the hyperbola. In this equation, however, although both x and y are in the first power, they are multiplied by each other, which practically makes a second power.

I have said that any equation containing x or y in the second power, and in no higher power, represents one of the curves of the conic sections whose type forms we have just given. But sometimes the equations do not correspond to these types exactly and require some manipulation to bring them into the type form.

Let us take the equation of a circle, namely, $x^2 + y^2 = 5^2$, and plot it as shown in Fig. 29.

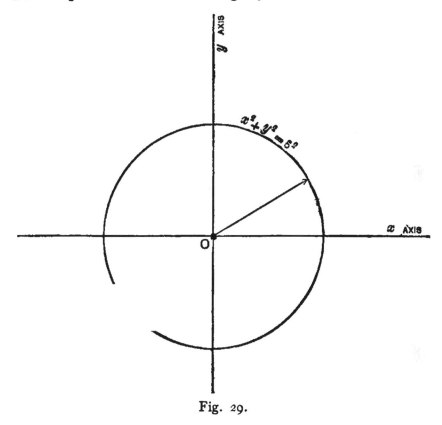

Fig. 29.

We see that it is a circle with its center at the intersection of the coördinate axes. Now take the equation $(x - 2)^2 + (y - 3)^2 = 5^2$. Plotting this, Fig. 30, we see that it is the same circle with its center at the point whose coördinates are 2 and 3. This equation and the first equation of the circle are identical in form, but frequently it is difficult at a glance to discover this identity, therefore much ingenuity is frequently required in detecting same.

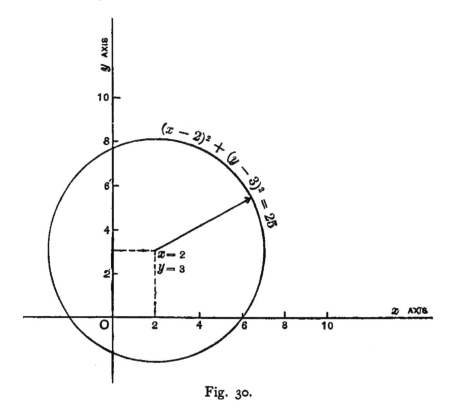

Fig. 30.

In plotting the equation of a hyperbola, $xy = 25$ (Fig. 31), we recognize this as a curve which is met with very frequently in engineering practice, and a knowledge of its general laws is of great value.

Similarly, in plotting a parabola (Fig. 32), $y^2 = 4x$, we see another familiar curve.

In this brief chapter we can only call attention to the conic sections, as their study is of academic more than

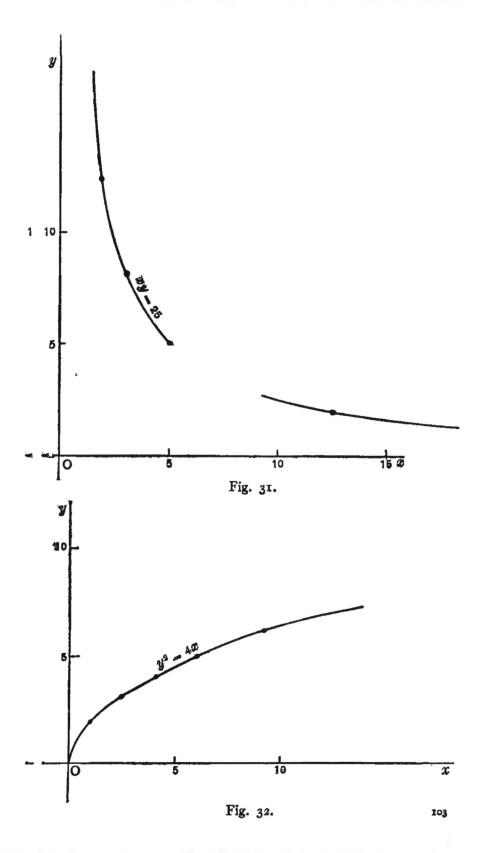

Fig. 31.

Fig. 32.

of pure engineering interest. However, as the student progresses in his knowledge of mathematics, I would suggest that he take up the subject in detail as one which will offer much fascination.

Other Curves. — All other equations containing unknown quantities which enter in h'gher powers than the second power, represent a large variety of curves called cubic curves.

The student may find the curve corresponding to engineering laws whose equations he will hereafter study. The main point of the whole discussion of this chapter is to teach him the methods of plotting, and if successful in this one point, this is as far as **we** shall go at the present time.

Intersection of Curves and Straight Lines. — When studying simultaneous equations we saw that if we had two equations showing the relation between two unknown quantities, such for instance as the equations

$$x + y = 7,$$
$$x - y = 3,$$

we could eliminate one of the unknown quantities in these equations and obtain the values of x and y which will satisfy both equations; thus, in the above equations, eliminating y, we have

$$2x = 10,$$
$$x = 5.$$

Substituting this value of x in one of the equations, we have
$$y = 2.$$

Now each one of the above equations represents a straight line, and each line can be plotted as shown in Fig. 33.

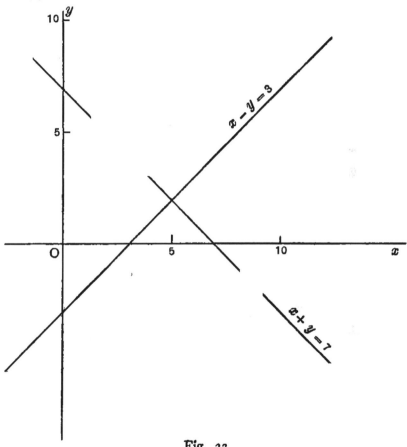

Fig. 33.

Their point of intersection is obviously a point on both lines. The coördinates of this point, then, $x = 5$ and $y = 2$, should satisfy both equations, and we have already seen this. Therefore, in general, where we have two equations each showing a relation in value between the two unknown quantities, x and y, by combining these equations, namely, eliminating one of the unknown quantities and solving for the other, our result will be the point or points of intersection of both curves represented by the equations. Thus, if we add the equations of two circles,

$$x^2 + y^2 = 4^2,$$
$$(x - 2)^2 + y^2 = 5^2,$$

and if the student plots these equations separately and then combines them, eliminating one of the unknown quantities and solving for the other, his results will be the points of intersection of both curves.

Plotting of Data. — When plotting mathematically with absolute accuracy the curve of an equation, whatever scale we use along one axis we must employ along the other axis. But, for practical results in plotting curves which show the relative values of several varying quantities during a test or which show the operation of machines under certain conditions, we depart from mathematical accuracy in the curve for the sake of convenience and choose such scales of value along each axis as we may deem appropriate. Thus, suppose we were plotting the characteristic curve of a shunt dynamo which had given the following sets of values from no load to full load operation:

VOLTS	AMPERES
122	0
120	5
118	10
116	15
114	19
111	22
107	25

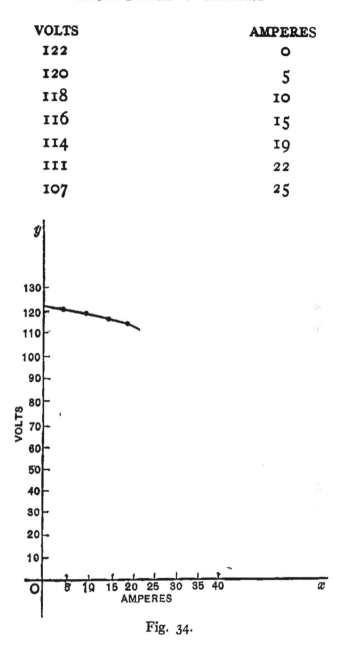

Fig. 34.

We plot this curve for convenience in a manner as shown in Fig. 34. Along the volts axis we choose a scale which is compressed to within one-half of the space that we choose for the amperes along the ampere axis. However, we might have chosen this entirely at our own discretion and the curve would have had the same significance to an engineer.

PROBLEMS

Plot the curves and lines corresponding to the following equations:

1. $x = 3y + 10$.

2. $2x + 5y = 15$.

3. $x - 2y = 4$.

4. $10y + 3x = -8$.

5. $x^2 + y^2 = 36$.

6. $x^2 = 16y$.

7. $x^2 - y^2 = 16$.

8. $3x^2 + (y - 2)^2 = 25$.

Find the intersections of the following curves and lines:

1. $3x + y = 10$,
   $4x - y = 6$.

2. $x^2 + y^2 = 81$,
   $x - y = 10$.

3. $\quad xy = 40$,
   $3x + y = 5$.

Plot the following volt-ampere curve:

VOLTS	AMPERES
550	0
548	20
545	39
541	55
536	79
529	91
521	102
510	115

CHAPTER XIV

ELEMENTARY PRINCIPLES OF THE CALCULUS

It is not my aim in this short chapter to do more than point out and explain a few of the fundamental ideas of the calculus which may be of value to a practical working knowledge of engineering. To the advanced student no study can offer more intellectual and to some extent practical interest than the advanced theories of calculus, but it must be admitted that very little beyond the fundamental principles ever enter into the work of the practical engineer.

In a general sense the study of calculus covers an investigation into the innermost properties of variable quantities, that is quantities which have variable values as against those which have absolutely constant, perpetual and absolutely fixed values. (In previous chapters we have seen what was meant by a *constant* quantity and what was meant by a *variable* quantity in an equation.) By the innermost properties of a *variable* quantity we mean finding out in the minutest detail just how this quantity originated; what infinitesimal (that is, exceedingly small) parts go to make it up; how it increases or diminishes with reference to other quantities; what its rate of increasing or diminishing may be; what its greatest and least values are; what is the smallest particle into which it may be divided; and what is the result of adding all of the smallest particles together. All of the processes of the calculus therefore are either *analysis* or *synthesis*, that is, either *tearing up* a quantity into its smallest parts or *building up* and *adding together* these smallest parts to make the quantity. We call the *analysis*, or tearing apart, *differentiation;* we call the *synthesis*, or building up, *integration.*

DIFFERENTIATION

Suppose we take the straight line (Fig. 35) of length *x*. If we divide it into a large number of parts, greater than a million or a billion or any number of which we

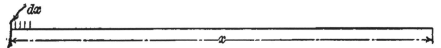

Fig. 35.

have any conception, we say that each part is infinitesimally small, — that is, it is small beyond conceivable length. We represent such inconceivably small lengths by an expression Δx or δx. Likewise, if we have a surface and divide it into infinitely small parts, and if we call *a* the area of the surface, the small infinitesimal portion of that surface we represent by Δa or δa. These quantities, namely, δx and δa, are called the *differential* of *x* and *a* respectively.

We have seen that the differential of a line of the length x is δx. Now suppose we have a square each of whose sides is x, as shown in Fig. 36. The area of that square is then x^2. Suppose now we increase the length

Fig. 36.

of each side by an infinitesimally small amount, δx, making the length of each side $x + \delta x$. If we complete a square with this new length as its side, the new square will obviously be larger than the old square by a very small amount. The actual area of the new square will be equal to the area of the old square + the additions to it. The area of the old square was equal to x^2. The addition consists of two fine strips each x long by δx wide and a small square having δx as the length of its side. The area of the addition then is

$$(x \times \delta x) + (x \times \delta x) + (\delta x \times \delta x) = \text{additional area.}$$

(The student should note this very carefully.) Therefore the addition equals

$$2\,x\,\delta x + (\delta x)^2 = \text{additional area.}$$

Now the smaller δx becomes, the smaller in more rapid proportion does δx^2, which is the area of the small square, become.   Likewise the smaller δx is, the thinner do the strips whose areas are $x\,\delta x$ become; but the strips do not diminish in value as fast as the small square diminishes, and, in fact, the small square vanishes so rapidly in comparison with the strips that even when the strips are of appreciable size the area of the small square is inappreciable, and we may say practically that by increasing the length of the side x of the square shown in Fig. 36 by the length δx we increase its area by the quantity $2\,x\,\delta x$.

Again, if we reduce the side x of the square by the length δx, we reduce the area of the square by the amount $2\,x\,\delta x$.   This infinitesimal quantity, out of a very large number of which the square consists or may be considered as made up of, is equal to the *differential* of the square, namely, the *differential* of x^2.   We thus see that the *differential* of the quantity x^2 is equal to $2\,x\,\delta x$. Likewise, if we had considered the case of a cube instead of a square, we would have found that the differential of the cube x^3 would have been $3\,x^2\,\delta x$.   Likewise, by more elaborate investigations we find that the differential of $x^4 = 4\,x^3\,\delta x$.   Summarizing, then, the foregoing results we have   differential of $x\ =\ \delta x$,

differential of $x^2\ =\ 2\,x\,\delta x$,

differential of $x^3\ =\ 3\,x^2\,\delta x$,

differential of $x^4\ =\ 4\,x^3\,\delta x$.

From these we see that there is a very simple and definite law by which we can at once find the differential of any power of x.

Law. — Reduce the power of x by one, multiply by δx and place before the whole a coefficient which is the same number as the power of x which we are differentiating; thus, if we differentiate x^5 we get $5\,x^4\,\delta x$; also, if we differentiate x^6 we get $6\,x^5\,\delta x$.

I will repeat here that it is necessary for the student to get a clear conception of what is meant by differentiation ; and I also repeat that in differentiating any quantity our object is to find out and get the value of the very small parts of which it is constructed (the rate of growth). Thus we have seen that a line is constructed of small lengths δx all placed together; that a square grows or evolves by placing fine strips one next the other; that a cube is built up of thin surfaces placed one over the other; and so on.

Differentiation Similar to Acceleration. — We have just said that finding the value of the *differential*, or one of the smallest particles whose gradual addition to a quantity makes the quantity, is the same as finding out *the rate of growth*, and this is what we understood by the ordinary term *acceleration* Now we can begin to see concretely just what we are aiming at in the term *differential*. The student should stop right here, think over all that has gone before and weigh each word of what we are saying with extreme care, for if he understands that the differentiation of a quantity gives us the rate of growth or acceleration of that quantity he has mastered the most important idea, in fact the *keynote idea* of all the calculus; I repeat, the *keynote idea*. Before going further let us stop for a little illustration.

Example. — If a train is running at a constant speed of ten miles an hour, the speed is constant, unvarying and therefore has no rate of change, since it does not change at all. If we call x the speed of the train, therefore x would be a constant quantity, and if we put it in an equation it would have a constant value and be called a *constant*. In algebra we have seen that we do not usually designate a constant or known quantity by the symbol x, but rather by the symbols a, k, etc.

Now on the other hand suppose the speed of the train was changing; say in the first hour it made ten miles, in the second hour eleven miles, in the third hour twelve miles, in the fourth hour thirteen miles, etc. It is evident that the speed is increasing one mile per hour each hour. This increase of speed we have always called the acceleration or rate of growth of the speed. Now if we designated the speed of the train by the symbol x, we see that x would be a variable quantity and would have a different value for every hour, every minute, every second, every instant that the train was running. The speed x would constantly at every instant have added to it a little more speed, namely δx, and if we can find the value of this small quantity δx for each instant of time we would have the *differential* of speed x, or in other words the *acceleration of the speed x*. Now let us repeat, x would have to be a *variable quantity* in order to have *any differential at all*, and if it is a variable quantity and has a *differential*, then that *differential* is the *rate of growth* or *acceleration* with which the value of that quantity x is increasing or diminishing as the case may be. We now see the significance of the term *differential*.

One more illustration. We all know that if a ball is thrown straight up in the air it starts up with great speed and gradually stops and begins to fall. Then as it falls it continues to increase its speed of falling until it strikes the earth with the same speed that it was thrown up with. Now we know that the force of gravity has been pulling on that ball from the time that it left our hands and has accelerated its speed backwards until it came to a stop in the air, and then speeded it to the earth. This instantaneous change in the speed of the ball we have called the acceleration of gravity, and is the rate of change of the speed of the ball. From careful observation we find this to be 32 ft. per second per second. A little further on we will learn how to express the concrete value of δx in simple form.

Differentiation of Constants.— Now let us remember that a *constant quantity*, since it has *no rate of change*, cannot be *differentiated;* therefore its differential is zero. If, however, a variable quantity such as x is multiplied by a constant quantity such as 6, making the quantity $6\,x$, of course this does not prevent you from differentiating the variable part, namely x; but of course the constant quantity remains unchanged; thus the differential of $6 = 0$.

But the differential of $3\,x\ -3\,\delta x$,
the differential of $4\,x^2 = 4$ times $2\,x\,\delta x = 8\,x\,\delta x$,
the differential of $2\,x^3 = 2$ times $3\,x^2\,\delta x = 6\,x^2\,\delta x$,
and so on.

Differential of a Sum or Difference. — We have seen how to find the *dffierential* of a single term.   Let us now take up an algebraic expression consisting of several terms with positive or negative signs before them; for example

$$x^2 - 2\,x + 6 + 3\,x^4.$$

In *differentiating* such an expression it is obvious that we must *differentiate* each term separately, for each term is separate and distinct from the other terms, and therefore its differential or rate of growth will be distinct and separate from the differential of the other terms; thus

The differential of $(x^2 - 2x + 6 + 3x^4)$

$$= 2x\,\delta x - 2\,\delta x + 12x^3\,\delta x.$$

We need scarcely say that if we differentiate one side of an algebraic equation we must also differentiate the other side; for we have already seen that whatever operation is performed to one side of an equation must be performed to the other side in order to retain the equality. Thus if we differentiate

$$x^2 + 4 = 6x - 10,$$

we get $\qquad 2x\,\delta x + o = 6\,\delta x - o,$

or $\qquad 2x\,\delta x = 6\,\delta x.$

Differentiation of a Product. — In Fig. 37 we have a rectangle whose sides are x and y and whose area is

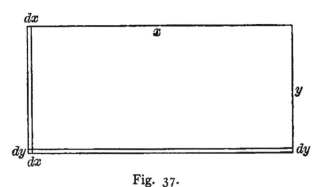

Fig. 37.

therefore equal to the product xy. Now increase its sides by a small amount and we have the old area added to by two thin strips and a small rectangle, thus:

New area = Old area $+ y\,\delta x + \delta y\,\delta x + x\,\delta y.$

$\delta y\ \delta x$ is negligibly small; therefore we see that the differential of the original area $xy = x\,\delta y + y\,\delta x$. This can be generalized for every case and we have the law

Law. — " The differential of the product of two variables is equal to the first multiplied by the differential of the second plus the second multiplied by the differential of the first." Thus,

$$\text{Differential } x^2y = x^2\,\delta y + 2\,yx\,\delta x.$$

This law holds for any number of variables.

$$\text{Differential } xyz = xy\,\delta z + xz\,\delta y + yz\,\delta x.$$

Differential of a Fraction. — If we are asked to differentiate the fraction $\dfrac{x}{y}$ we first write it in the form xy^{-1}, using the negative exponent; now on differentiating we have

$$\text{Differential } xy^{-1} = -\,xy^{-2}\,\delta y + y^{\,1}\,\delta x$$
$$= -\,\frac{x\,\delta y}{y^2} + \frac{\delta x}{y}.$$

Reducing to a common denominator we have

$$\text{Differential } xy^{-1} \text{ or } \frac{x}{y} - -\,\frac{x\,\delta y}{y^2} + \frac{y\,\delta x}{y^2}$$
$$= \frac{y\,\delta x - x\,\delta y}{y^2}.$$

Law. — The differential of a fraction is then seen to be equal to the differential of the numerator times the denominator, minus the differential of the denominator times the numerator, all divided by the square of the denominator.

Differential of One Quantity with Respect to Another. — Thus far we have considered the differential of a *variable* with respect to itself, that is, we have considered its rate of development in so far as it was itself alone concerned. Suppose however we have two variable quantities dependent on each other, that is, as one changes the other changes, and we are asked to find the rate of change of the one with respect to the other, that is, to find the rate of change of one knowing the rate of change of the other. At a glance we see that this should be a very simple process, for if we know the relation which subsists between two variable quantities, this relation being expressed in the form of an equation between the two quantities, we should readily be able to tell the relation which will hold between similar deductions from these quantities. Let us for instance take the equation

$$x = y + 2.$$

Here we have the two variables x and y tied together by an equation which establishes a relation between them. As we have previously seen, if we give any definite value to y we will find a corresponding value for x. Referring to our chapter on coördinate geometry we see that this is the equation of the line shown in Fig. 38.

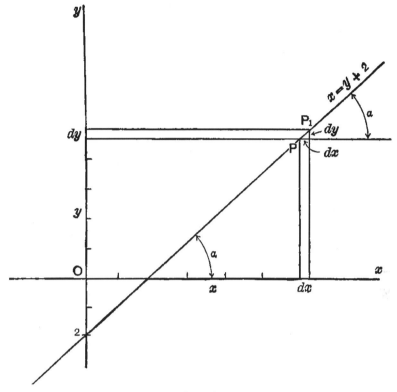

Fig. 38.

Let us take any point P on this line. Its coördinates are y and x respectively. Now choose another point P_1 a short distance away from P on the same line. The *abscissa* of this new point will be a little longer than that of the old point, and will equal $x + \delta x$, while the *ordinate* y of the old point has been increased by δy, making the *ordinate* of the new point $y + \delta y$.

From Fig. 38 we see that

$$\tan \alpha = \frac{\delta y}{\delta x}.$$

Therefore, if we know the tangent α and know either δy or δx we can find the other.

In this example our equation represents a straight line, but the same would be true for any curve represented by any equation between x and y no matter how complicated; thus Fig. 39 shows the relation between

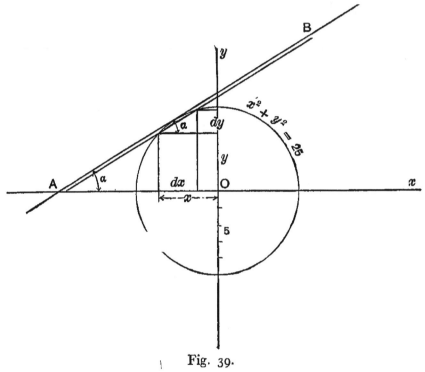

Fig. 39.

δx and δy at one point of the curve (a circle) whose equation is $x^2 + y^2 = 25$. For every other point of the circle $\tan \alpha$ or $\dfrac{\delta y}{\delta x}$ will have a different value. δx and δy while shown quite large in the figure for demonstration's sake are inconceivably small in reality; therefore the line AB in the figure is really a tangent of the curve, and $\not< \alpha$ is the angle which it makes with the x axis. For every point on the curve this angle will be different.

Mediate Differentiation. — Summarizing the foregoing we see that if we know any two of the three unknowns in equation $\tan \alpha = \dfrac{\delta y}{\delta x}$ we can find the third.

Some textbooks represent $\tan \alpha$ or $\dfrac{\delta y}{\delta x}$ by y_x and, $\dfrac{\delta x}{\delta y}$ by x_y. This is a convenient notation and we will use it here. Therefore we have

$$\delta x \tan \alpha = \delta y,$$

$$\frac{\delta y}{\tan \alpha} - \delta x$$

or
$$\delta y = \delta x \, y_x,$$

$$\delta x = \delta y \, x_y.$$

This shows us that if we differentiate the quantity $3 x^2$ as to x we obtain $6 x \, \delta x$, but if we had wished to differentiate it with respect to y we would first have to differentiate it with respect to x and then multiply by x_y thus:

Differentiation of $3 x^2$ as to $y = 6 x \, \delta y \, x_y$.

Likewise if we have $4 y^3$ and we wish to differentiate it with respect to x we have

Differential of $4 y^3$ as to $x = 12 y^2 \, \delta x \, y_x.$

This is called *mediate differentiation* and is resorted to primarily because we can differentiate a power with respect to itself readily, but not with respect to some other variable.

Law. — To differentiate any expression containing x as to y, first differentiate it as to x and then multiply by x_y δy or vice versa.

We need this principle if we find the differential of several terms some containing x and some y; thus if we differentiate the equation $2\,x^2 = y^3 - 10$ with respect to x we get

$$4\,x\,\delta x = 3\,y^2 y_x\,\delta x + 0,$$

or
$$4\,x = 3\,y^2 y_x,$$

or
$$y_x = \frac{4\,x}{3\,y^2},$$

or
$$\tan \alpha = \frac{4\,x}{3\,y^2}.$$

From this we see that by differentiating the original equation of the curve we got finally an equation giving the value $\tan \alpha$ in terms of x and y, and if we fill out the exact numerical values of x and y for any particular point of the curve we will immediately be able to determine the slant of the tangent of the curve at this point, as we will numerically have the value of tangent α, and α is the angle that the tangent makes with the x axis.

In just the same manner that we have proceeded here we can proceed to find the direction of the tangent of any curve whose equation we know. The differential of y as to x, namely $\dfrac{\delta y}{\delta x}$ or y_x, must be kept in mind as the rate of change of y with respect to x, and nothing so vividly portrays this fact as the inclination of the tangent to the curve which shows the bend of the curve at every point.

Differentials of Other Functions. — By elaborate processes which cannot be mentioned here we find that the

Differential of the sine x as to $x = $ cosine $x\ \delta x$.

Differential of the cosine x as to $x = -$ sin $x\delta x$.

Differential of the log x as to $x - \dfrac{1}{x}\ \delta x$.

Differential of the sine y as to $x = $ cosine $y\ y_x\ \delta x$.

Differential of the cosine y as to $x = -$sine $y\ y_x\ \delta x$.

Differential of the log y as to $x = \dfrac{1}{y} y_x\ \delta x$.

Maxima and Minima. — Referring back to the circle, Fig. 39, once more, we see that

$$x^2 + y^2 = 25.$$

Differentiating this equation with reference to x we have

$$2\ x\ \delta x + 2\ y\ y_x\ \delta x = 0,$$

or

$$2\ x + 2\ y\ y_x = 0,$$

or

$$y_x = -\frac{x}{y}$$

Therefore

$$\tan \alpha - \frac{x}{y}$$

Now when $\tan \alpha = 0$ it is evident that the tangent to the curve is parallel to the x axis. At this point y is either a maximum or a minimum which can be readily determined on reference to the curve.

$$o - \frac{x}{y},$$

$$x = o.$$

Therefore $x = o$ when y is maximum and in this particular curve also minimum.

Law. — If we want to find the maximum or minimum value of x in any equation containing x and y, we differentiate the equation with reference to y and solve for the value of x_y; this we make equal to o and then we solve for the value of y in the resulting equation.

Example. — Find the maximum or minimum value of x in the equation

$$y^2 = 14\, x.$$

Differentiating with respect to y we have

$$2\, y\, \delta y = 14\, x_y\, \delta y,$$

$$x_y = \frac{2\, y}{14}.$$

Equating this to o we have

$$\frac{2\, y}{14} = o,$$

or $$y = o.$$

In other words, we find that x has its minimum value

when $y = 0$. We can readily see that this is actually the case in Fig. 40, which shows the curve (a parabola).

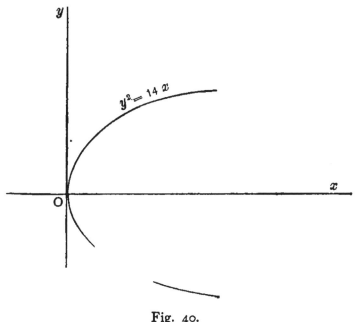

$$y^2 = 14\,x$$

Fig. 40.

INTEGRATION

Integration is the exact opposite of differentiation. In differentiation we divide a body into its constituent parts, in integration we add these constituent parts together to produce the body.

Integration is indicated by the sign $\int$; thus, if we wished to integrate δx we would write

$$\int \delta x.$$

Since integration is the opposite of differentiation, if we are given a quantity and asked to integrate it, our answer would be that quantity which differentiated will give us our original quantity. For example, we detect δx as the derivative of x; therefore the integral $\int \delta x = x$. Likewise, we detect $4\ x^3\ \delta x$ as the differential of x^4; therefore the integral $\int 4\ x^3\ \delta x = x^4$

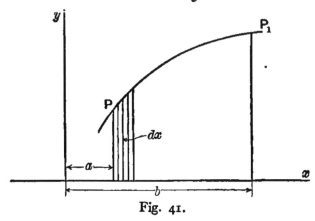

Fig. 41.

If we consider the line AB (Fig. 35) to be made up of small parts δx, we could sum up these parts thus:

$$\delta x + \delta x + \delta x + \delta x + \delta x + \delta x \ \ldots \ldots$$

for millions of parts. But integration enables us to express this more simply and $\int \delta x$ means the summation of every single part δx which goes to make up the line AB, no matter how many parts there may be or how small each part. But x is the whole length of the line of indefinite length. To sum up any portion of the line between the points or limits $x = 1$ and $x = 4$, we would write

$$\int_{x=1}^{x=4} \delta x \quad - (x) \quad \begin{array}{c} x = 4 \\ x = 1 \end{array}$$

Now these are definite integrals because they indicate exactly between what limits or points we wish to find the length of the line. This is true for all integrals. Where no limits of integration are shown the integral will yield only a general result, but when limits are stated between which summation is to be made, then we have a definite integral whose precise value we may ascertain.

Refer back to the expression $(x)\begin{smallmatrix} x = 4 \\ x = 1 \end{smallmatrix}$; in order to solve this, substitute inside of the parenthesis the value of x for the upper limit of x, namely, 4, and substitute and subtract the value of x at the lower limit, namely, 1; we then get

$$(x)\begin{smallmatrix} x = 4 \\ \\ x = 1 \end{smallmatrix} - (4 - 1) = 3.$$

Thus 3 is the length of the line between 1 and 4. Or, to give another illustration, suppose the solution of some integral had given us

$$(x^2 - 1)\begin{smallmatrix} x = 3 \\ x = 2 \end{smallmatrix},$$

then

$$(x^2 - 1)\begin{smallmatrix} x = 3 \\ \\ x = 2 \end{smallmatrix} - (3^2 - 1) - (2^2 - 1) = 5.$$

Here we simply substituted for x in the parenthesis its upper limit, then subtracted from the quantity thus obtained another quantity, which is had by substituting the lower limit of x.

By higher mathematics and the theories of series we prove that the integral of any power of a variable as to itself is obtained by increasing the exponent by one and dividing by the new exponent, thus:

$$\int x^2 \, \delta x - \frac{x^3}{3},$$

$$\int 4 \, x^5 \, \delta x = \frac{4 \, x^6}{6}.$$

On close inspection this is seen to be the inverse of the law of differentiation, which says to decrease the exponent by one and multiply by the old exponent.

So many and so complex are the laws of nature and so few and so limited the present conceptions of man that only a few type forms of integrals may be actually integrated. If the quantity under the integral sign by some manipulation or device is brought into a form where it is recognized as the differential of another quantity, then integrating it will give that quantity.

The Integral of an Expression. — The integral of an algebraic expression consisting of several terms is equal to the sum of the integrals of each of the separate terms; thus,

$$\int x^2 \, \delta x + 2 \, x \, \delta x + 3 \, \delta x$$

is the same thing as

$$\int x^2 \, \delta x + \int 2 \, x \, \delta x + \int 3 \, \delta x.$$

The most common integrals to be met with practically are:

(1) The integrals of some power of the variable whose solution we have just explained

(2) The integrals of the sine and cosine, which are

$$\int \text{cosine } x \, \delta x = \text{sine } x,$$

$$\int \text{sine } x \, \delta x = - \text{cosine } x.$$

(3) The integral of the reciprocal, which is

$$\int \frac{1}{x} \delta x = \log_\bullet x.^*$$

Areas. — Up to the present we have considered only the integration of a quantity with respect to itself. Suppose now we integrate one quantity with respect to another.

In Fig. 41 we have the curve PP_1, which is the graphical representation of some equation containing x and y. If we wish to find the area which lies between the curve and the x axis and between the two vertical lines drawn at distances $x = a$ and $x = b$ respectively, we divide the space up by vertical lines drawn δx distance apart. Now we would have a large number of small strips each δx wide and all having different heights, namely, y_1, y_2, y_3, y_4, etc.

The enumeration of all these areas would then be

$$y_1 \, \delta x + y_2 \, \delta x + y_3 \, \delta x + y_4 \, \delta x, \text{ etc.}$$

* Log$_\bullet$ means natural logarithm or logarithm to the Napierian base e which is equal to 2.718 as distinguished from ordinary logarithms to the base 10. In fact wherever log appears in this chapter it means log$_\bullet$.

Now calculus enables us to say

$$\text{Area wanted} = \int_{x=a}^{x=b} y \, \delta x.$$

This integral $\int_{x=a}^{x=b} y \, \delta x$ cannot be readily solved. If it were $\int x \, \delta x$ we have seen that the result would be $\dfrac{x^2}{2}$; but this is not the case with $\int y \, \delta x$. We must then find some way to replace y in this integral by some expression containing x. It is here then that we have to resort to the equation of the curve PP_1. From this equation we find the value of y in terms of x; we then substitute this value of y in the integral $\int y \, \delta x$, and then having an integral of x as to itself we can readily solve it. Now, if the equation of the curve PP_1 is a complex one this process becomes very difficult and sometimes impossible.

A simple case of the above is the hyperbola $xy = 10$ (Fig. 42). If we wish to get the value of the shaded area we have

$$\text{Shaded area} = \int_{x=5\text{ ft.}}^{x=12\text{ ft.}} y \, \delta x.$$

From the equation of this curve we have

$$xy = 10,$$

$$y = \frac{10}{x}. \; \therefore$$

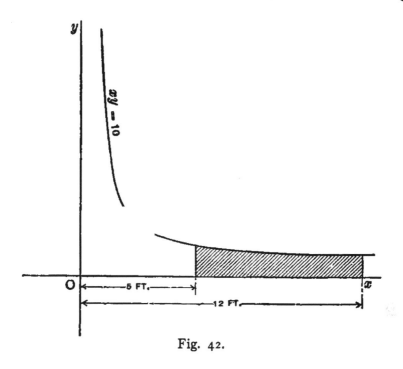

Fig. 42.

Therefore, substituting we have

$$\text{Shaded area} = \int_{x=5}^{x=12} \frac{10}{x} \, \delta x.$$

$$\text{Area} = 10 \, (\log_e x) \quad \begin{array}{l} x = 12 \\ x = 5 \end{array}$$

$$- 10 \, (\log_e 12 - \log_e 5)$$

$$- 10 \, (2.4817 - 1.6077).$$

$$\text{Area} = 8.740 \text{ sq. ft.}$$

Beyond this brief gist of the principles of calculus we can go no further in this chapter. The student may not understand the theories herein treated of at first — in fact, it will take him, as it has taken every student, many months before the true conceptions of calculus dawn on him clearly. And, moreover, it is not essential that he know calculus at all to follow the ordinary engineering discussions. It is only where a student wishes to obtain the deepest insight into the science that he needs calculus, and to such a student I hope this chapter will be of service as a brief preliminary to the difficulties and complexities of that subject.

PROBLEMS

1. Differentiate $2\,x^3$ as to x.
2. Differentiate $12\,x^2$ as to x.
3. Differentiate $8\,x^5$ as to x.
4. Differentiate $3\,x^2 + 4\,x + 10 = 5\,x^3$ as to x.
5. Differentiate $4\,y^2 - 3\,x$ as to y.
6. Differentiate $14\,y^4x^3$ as to y.
7. Differentiate $\dfrac{x^2}{y}$ as to x.
8. Differentiate $2\,y^2 - 4\,qx$ as to y.

Find y_x in the following equations:

9. $x^2 + 2\,y^2 = 100$.
10. $x^3 + y = 5$.
11. $x^2 - y^2 = 25$.
12. $5\,xy = 12$.

13. What angle does the tangent line to the circle $x^2 + y^2 = 9$ make with the x axis at the point where $x = 2$?

14. What is the minimum value of y in the equation $x^2 = 15\,y$?

15. Solve $\int 2\,x^3\,\delta x$.

16. Solve $\int 5\,x^2\,\delta x$.

17. Solve $\int 10\,ax\,\delta x + 5\,x^2\,\delta x + 3\,\delta x$.

18. Solve $\int 3\ \text{sine}\ x\,\delta x$.

19. Solve $\int 2\ \text{cosine}\ x\,\delta x$.

20. Solve $\int_{x=2}^{x=5} 3\,x^2\,\delta x$.

21. Solve $\int_{x=2}^{x=18} y\,\delta x$ if $xy = 4$.

22. Differentiate 10 sine x as to x.

23. Differentiate cosine x sine x as to x.

24. Differentiate log x as to x.

25. Differentiate $\dfrac{y^2}{x^2}$ as to x.

The following tables are reproduced from Ames and Bliss's "Manual of Experimental Physics" by permission of the American Book Company.

	0	1	2	3	4	5	6	7	8	9	1 2 3	4 5 6	7 8 9
10	0000	0043	0086	0128	0170	0212	0253	0294	0334	0374	Use preceding Table		
11	0414	0453	0492	0531	0569	0607	0645	0682	0719	0755	4 8 11	15 19 23	26 30 34
12	0792	0828	0864	0899	0934	0969	1004	1038	1072	1106	3 7 10	14 17 21	24 28 31
13	1139	1173	1206	1239	1271	1303	1335	1367	1399	1430	3 6 10	13 16 19	23 26 29
14	1461	1492	1523	1553	1584	1614	1644	1673	1703	1732	3 6 9	12 15 18	21 24 27
15	1761	1790	1818	1847	1875	1903	1931	1959	1987	2014	3 6 8	11 14 17	20 22 25
16	2041	2068	2095	2122	2148	2175	2201	2227	2253	2279	3 5 8	11 13 16	18 21 24
17	2304	2330	2355	2380	2405	2430	2455	2480	2504	2529	2 5 7	10 12 15	17 20 22
18	2553	2577	2601	2625	2648	2672	2695	2718	2742	2765	2 5 7	9 12 14	16 19 21
19	2788	2810	2833	2856	2878	2900	2923	2945	2967	2989	2 4 7	9 11 13	16 18 20
20	3010	3032	3054	3075	3096	3118	3139	3160	3181	3201	2 4 6	8 11 13	15 17 19
21	3222	3243	3263	3284	3304	3324	3345	3365	3385	3404	2 4 6	8 10 12	14 16 18
22	3424	3444	3464	3483	3502	3522	3541	3560	3579	3598	2 4 6	8 10 12	14 15 17
23	3617	3636	3655	3674	3692	3711	3729	3747	3766	3784	2 4 6	7 9 11	13 15 17
24	3802	3820	3838	3856	3874	3892	3909	3927	3945	3962	2 4 5	7 9 11	12 14 16
25	3979	3997	4014	4031	4048	4065	4082	4099	4116	4133	2 3 5	7 9 10	12 14 15
26	4150	4166	4183	4200	4216	4232	4249	4265	4281	4298	2 3 5	7 8 10	11 13 15
27	4314	4330	4346	4362	4378	4393	4409	4425	4440	4456	2 3 5	6 8 9	11 13 14
28	4472	4487	4502	4518	4533	4548	4564	4579	4594	4609	2 3 5	6 8 9	11 12 14
29	4624	4639	4654	4669	4683	4698	4713	4728	4742	4757	1 3 4	6 7 9	10 12 13
30	4771	4786	4800	4814	4829	4843	4857	4871	4886	4900	1 3 4	6 7 9	10 11 13
31	4914	4928	4942	4955	4969	4983	4997	5011	5024	5038	1 3 4	6 7 8	10 11 12
32	5051	5065	5079	5092	5105	5119	5132	5145	5159	5172	1 3 4	5 7 8	9 11 12
33	5185	5198	5211	5224	5237	5250	5263	5276	5289	5302	1 3 4	5 6 8	9 10 12
34	5315	5328	5340	5353	5366	5378	5391	5403	5416	5428	1 3 4	5 6 8	9 10 11
35	5441	5453	5465	5478	5490	5502	5514	5527	5539	5551	1 2 4	5 6 7	9 10 11
36	5563	5575	5587	5599	5611	5623	5635	5647	5658	5670	1 2 4	5 6 7	8 10 11
37	5682	5694	5705	5717	5729	5740	5752	5763	5775	5786	1 2 3	5 6 7	8 9 10
38	5798	5809	5821	5832	5843	5855	5866	5877	5888	5899	1 2 3	5 6 7	8 9 10
39	5911	5922	5933	5944	5955	5966	5977	5988	5999	6010	1 2 3	4 5 7	8 9 10
40	6021	6031	6042	6053	6064	6075	6085	6096	6107	6117	1 2 3	4 5 6	8 9 10
41	6128	6138	6149	6160	6170	6180	6191	6201	6212	6222	1 2 3	4 5 6	7 8 9
42	6232	6243	6253	6263	6274	6284	6294	6304	6314	6325	1 2 3	4 5 6	7 8 9
43	6335	6345	6355	6365	6375	6385	6395	6405	6415	6425	1 2 3	4 5 6	7 8 9
44	6435	6444	6454	6464	6474	6484	6493	6503	6513	6522	1 2 3	4 5 6	7 8 9
45	6532	6542	6551	6561	6571	6580	6590	6599	6609	6618	1 2 3	4 5 6	7 8 9
46	6628	6637	6646	6656	6665	6675	6684	6693	6702	6712	1 2 3	4 5 6	7 7 8
47	6721	6730	6739	6749	6758	6767	6776	6785	6794	6803	1 2 3	4 5 5	6 7 8
48	6812	6821	6830	6839	6848	6857	6866	6875	6884	6893	1 2 3	4 4 5	6 7 8
49	6902	6911	6920	6928	6937	6946	6955	6964	6972	6981	1 2 3	4 4 5	6 7 8
50	6990	6998	7007	7016	7024	7033	7042	7050	7059	7067	1 2 3	3 4 5	6 7 8
51	7076	7084	7093	7101	7110	7118	7126	7135	7143	7152	1 2 3	3 4 5	6 7 8
52	7160	7168	7177	7185	7193	7202	7210	7218	7226	7235	1 2 2	3 4 5	6 7 7
53	7243	7251	7259	7267	7275	7284	7292	7300	7308	7316	1 2 2	3 4 5	6 6 7
54	7324	7332	7340	7348	7356	7364	7372	7380	7388	7396	1 2 2	3 4 5	6 6 7

	0	1	2	3	4	5	6	7	8	9	1 2 3	4 5 6	7 8 9
55	7404	7412	7419	7427	7435	7443	7451	7459	7466	7474	1 2 2	3 4 5	5 6 7
56	7482	7490	7497	7505	7513	7520	7528	7536	7543	7551	1 2 2	3 4 5	5 6 7
57	7559	7566	7574	7582	7589	7597	7604	7612	7619	7627	1 2 2	3 4 5	5 6 7
58	7634	7642	7649	7657	7664	7672	7679	7686	7694	7701	1 1 2	3 4 4	5 6 7
59	7709	7716	7723	7731	7738	7745	7752	7760	7767	7774	1 1 2	3 4 4	5 6 7
60	7782	7789	7796	7803	7810	7818	7825	7832	7839	7846	1 1 2	3 4 4	5 6 6
61	7853	7860	7868	7875	7882	7889	7896	7903	7910	7917	1 1 2	3 4 4	5 6 6
62	7924	7931	7938	7945	7952	7959	7966	7973	7980	7987	1 1 2	3 3 4	5 6 6
63	7993	8000	8007	8014	8021	8028	8035	8041	8048	8055	1 1 2	3 3 4	5 5 6
64	8062	8069	8075	8082	8089	8096	8102	8109	8116	8122	1 1 2	3 3 4	5 5 6
65	8129	8136	8142	8149	8156	8162	8169	8176	8182	8189	1 1 2	3 3 4	5 5 6
66	8195	8202	8209	8215	8222	8228	8235	8241	8248	8254	1 1 2	3 3 4	5 5 6
67	8261	8267	8274	8280	8287	8293	8299	8306	8312	8319	1 1 2	3 3 4	5 5 6
68	8325	8331	8338	8344	8351	8357	8363	8370	8376	8382	1 1 2	3 3 4	4 5 6
69	8388	8395	8401	8407	8414	8420	8426	8432	8439	8445	1 1 2	2 3 4	4 5 6
70	8451	8457	8463	8470	8476	8482	8488	8494	8500	8506	1 1 2	2 3 4	4 5 6
71	8513	8519	8525	8531	8537	8543	8549	8555	8561	8567	1 1 2	2 3 4	4 5 5
72	8573	8579	8585	8591	8597	8603	8609	8615	8621	8627	1 1 2	2 3 4	4 5 5
73	8633	8639	8645	8651	8657	8663	8669	8675	8681	8686	1 1 2	2 3 4	4 5 5
74	8692	8698	8704	8710	8716	8722	8727	8733	8739	8745	1 1 2	2 3 4	4 5 5
75	8751	8756	8762	8768	8774	8779	8785	8791	8797	8802	1 1 2	2 3 3	4 5 5
76	8808	8814	8820	8825	8831	8837	8842	8848	8854	8859	1 1 2	2 3 3	4 5 5
77	8865	8871	8876	8882	8887	8893	8899	8904	8910	8915	1 1 2	2 3 3	4 4 5
78	8921	8927	8932	8938	8943	8949	8954	8960	8965	8971	1 1 2	2 3 3	4 4 5
79	8976	8982	8987	8993	8998	9004	9009	9015	9020	9025	1 1 2	2 3 3	4 4 5
80	9031	9036	9042	9047	9053	9058	9063	9069	9074	9079	1 1 2	2 3 3	4 4 5
81	9085	9090	9096	9101	9106	9112	9117	9122	9128	9133	1 1 2	2 3 3	4 4 5
82	9138	9143	9149	9154	9159	9165	9170	9175	9180	9186	1 1 2	2 3 3	4 4 5
83	9191	9196	9201	9206	9212	9217	9222	9227	9232	9238	1 1 2	2 3 3	4 4 5
84	9243	9248	9253	9258	9263	9269	9274	9279	9284	9289	1 1 2	2 3 3	4 4 5
85	9294	9299	9304	9309	9315	9320	9325	9330	9335	9340	1 1 2	2 3 3	4 4 5
86	9345	9350	9355	9360	9365	9370	9375	9380	9385	9390	1 1 2	2 3 3	4 4 5
87	9395	9400	9405	9410	9415	9420	9425	9430	9435	9440	0 1 1	2 2 3	3 4 4
88	9445	9450	9455	9460	9465	9469	9474	9479	9484	9489	0 1 1	2 2 3	3 4 4
89	9494	9499	9504	9509	9513	9518	9523	9528	9533	9538	0 1 1	2 2 3	3 4 4
90	9542	9547	9552	9557	9562	9566	9571	9576	9581	9586	0 1 1	2 2 3	3 4 4
91	9590	9595	9600	9605	9609	9614	9619	9624	9628	9633	0 1 1	2 2 3	3 4 4
92	9638	9643	9647	9652	9657	9661	9666	9671	9675	9680	0 1 1	2 2 3	3 4 4
93	9685	9689	9694	9699	9703	9708	9713	9717	9722	9727	0 1 1	2 2 3	3 4 4
94	9731	9736	9741	9745	9750	9754	9759	9763	9768	9773	0 1 1	2 2 3	3 4 4
95	9777	9782	9786	9791	9795	9800	9805	9809	9814	9818	0 1 1	2 2 3	3 4 4
96	9823	9827	9832	9836	9841	9845	9850	9854	9859	9863	0 1 1	2 2 3	3 4 4
97	9868	9872	9877	9881	9886	9890	9894	9899	9903	9908	0 1 1	2 2 3	3 4 4
98	9912	9917	9921	9926	9930	9934	9939	9943	9948	9952	0 1 1	2 2 3	3 4 4
99	9956	9961	9965	9969	9974	9978	9983	9987	9991	9996	0 1 1	2 2 3	3 3 4

	0′	6′	12′	18′	24′	30′	36′	42′	48′	54′	1	2	3	4	5
0°	0000	0017	0035	0052	0070	0087	0105	0122	0140	0157	3	6	9	12	15
1	0175	0192	0209	0227	0244	0262	0279	0297	0314	0332	3	6	9	12	15
2	0349	0366	0384	0401	0419	0436	0454	0471	0488	0506	3	6	9	12	15
3	0523	0541	0558	0576	0593	0610	0628	0645	0663	0680	3	6	9	12	15
4	0698	0715	0732	0750	0767	0785	0802	0819	0837	0854	3	6	9	12	15
5	0872	0889	0906	0924	0941	0958	0976	0993	1011	1028	3	6	9	12	14
6	1045	1063	1080	1097	1115	1132	1149	1167	1184	1201	3	6	9	12	14
7	1219	1236	1253	1271	1288	1305	1323	1340	1357	1374	3	6	9	12	14
8	1392	1409	1426	1444	1461	1478	1495	1513	1530	1547	3	6	9	12	14
9	1564	1582	1599	1616	1633	1650	1668	1685	1702	1719	3	6	9	12	14
10	1736	1754	1771	1788	1805	1822	1840	1857	1874	1891	3	6	9	12	14
11	1908	1925	1942	1959	1977	1994	2011	2028	2045	2062	3	6	9	11	14
12	2079	2096	2113	2130	2147	2164	2181	2198	2215	2232	3	6	9	11	14
13	2250	2267	2284	2300	2317	2334	2351	2368	2385	2402	3	6	8	11	14
14	2419	2436	2453	2470	2487	2504	2521	2538	2554	2571	3	6	8	11	14
15	2588	2605	2622	2639	2656	2672	2689	2706	2723	2740	3	6	8	11	14
16	2756	2773	2790	2807	2823	2840	2857	2874	2890	2907	3	6	8	11	14
17	2924	2940	2957	2974	2990	3007	3024	3040	3057	3074	3	6	8	11	14
18	3090	3107	3123	3140	3156	3173	3190	3206	3223	3239	3	6	8	11	14
19	3256	3272	3289	3305	3322	3338	3355	3371	3387	3404	3	5	8	11	14
20	3420	3437	3453	3469	3486	3502	3518	3535	3551	3567	3	5	8	11	14
21	3584	3600	3616	3633	3649	3665	3681	3697	3714	3730	3	5	8	11	14
22	3746	3762	3778	3795	3811	3827	3843	3859	3875	3891	3	5	8	11	14
23	3907	3923	3939	3955	3971	3987	4003	4019	4035	4051	3	5	8	11	14
24	4067	4083	4099	4115	4131	4147	4163	4179	4195	4210	3	5	8	11	13
25	4226	4242	4258	4274	4289	4305	4321	4337	4352	4368	3	5	8	11	13
26	4384	4399	4415	4431	4446	4462	4478	4493	4509	4524	3	5	8	10	13
27	4540	4555	4571	4586	4602	4617	4633	4648	4664	4679	3	5	8	10	13
28	4695	4710	4726	4741	4756	4772	4787	4802	4818	4833	3	5	8	10	13
29	4848	4863	4879	4894	4909	4924	4939	4955	4970	4985	3	5	8	10	13
30	5000	5015	5030	5045	5060	5075	5090	5105	5120	5135	3	5	8	10	13
31	5150	5165	5180	5195	5210	5225	5240	5255	5270	5284	2	5	7	10	12
32	5299	5314	5329	5344	5358	5373	5388	5402	5417	5432	2	5	7	10	12
33	5446	5461	5476	5490	5505	5519	5534	5548	5563	5577	2	5	7	10	12
34	5592	5606	5621	5635	5650	5664	5678	5693	5707	5721	2	5	7	10	12
35	5736	5750	5764	5779	5793	5807	5821	5835	5850	5864	2	5	7	10	12
36	5878	5892	5906	5920	5934	5948	5962	5976	5990	6004	2	5	7	9	12
37	6018	6032	6046	6060	6074	6088	6101	6115	6129	6143	2	5	7	9	12
38	6157	6170	6184	6198	6211	6225	6239	6252	6266	6280	2	5	7	9	11
39	6293	6307	6320	6334	6347	6361	6374	6388	6401	6414	2	4	7	9	11
40	6428	6441	6455	6468	6481	6494	6508	6521	6534	6547	2	4	7	9	11
41	6561	6574	6587	6600	6613	6626	6639	6652	6665	6678	2	4	7	9	11
42	6691	6704	6717	6730	6743	6756	6769	6782	6794	6807	2	4	6	9	11
43	6820	6833	6845	6858	6871	6884	6896	6909	6921	6934	2	4	6	8	11
44	6947	6959	6972	6984	6997	7009	7022	7034	7046	7059	2	4	6	8	10

7254	7266
7373	7385
7490	7501
7604	7615
7716	7727
7826	7837
7934	7944
8039	8049
8141	8151
8241	8251
8339	8348
8434	8443
8526	8536
8616	8625
8704	8712
8788	8796
8870	8878
8949	8957
9026	9033
9100	9107

9367	9373
9426	9432
9483	9489
9537	9542
9588	9593
9636	9641
9681	9686
9724	9728
9763	9767
9799	9803
9833	9836
9863	9866
9890	9893
9914	9917
9936	9938
9954	9956
9969	9971
9981	9982
9990	9991
9997	9997

	0′	6′	12′	18′	24′	30′	36′	42′	48′	54′	1	2	3	4	5
0°	1.000	1.000 nearly	1.000 nearly	1.000 nearly	1.000 nearly	9999	9999	9999	9999	9999	0	0	0	0	0
1	9998	9998	9998	9997	9997	9997	9996	9996	9995	9995	0	0	0	0	0
2	9994	9993	9993	9992	9991	9990	9990	9989	9988	9987	0	0	0	1	1
3	9986	9985	9984	9983	9982	9981	9980	9979	9978	9977	0	0	1	1	1
4	9976	9974	9973	9972	9971	9969	9968	9966	9965	9963	0	0	1	1	1
5	9962	9960	9959	9957	9956	9954	9952	9951	9949	9947	0	1	1	1	2
6	9945	9943	9942	9940	9938	9936	9934	9932	9930	9928	0	1	1	1	2
7	9925	9923	9921	9919	9917	9914	9912	9910	9907	9905	0	1	1	2	2
8	9903	9900	9898	9895	9893	9890	9888	9885	9882	9880	0	1	1	2	2
9	9877	9874	9871	9869	9866	9863	9860	9857	9854	9851	0	1	1	2	2
10	9848	9845	9842	9839	9836	9833	9829	9826	9823	9820	1	1	2	2	3
11	9816	9813	9810	9806	9803	9799	9796	9792	9789	9785	1	1	2	2	3
12	9781	9778	9774	9770	9767	9763	9759	9755	9751	9748	1	1	2	3	3
13	9744	9740	9736	9732	9728	9724	9720	9715	9711	9707	1	1	2	3	3
14	9703	9699	9694	9690	9686	9681	9677	9673	9668	9664	1	1	2	3	4
15	9659	9655	9650	9646	9641	9636	9632	9627	9622	9617	1	2	2	3	4
16	9613	9608	9603	9598	9593	9588	9583	9578	9573	9568	1	2	2	3	4
17	9563	9558	9553	9548	9542	9537	9532	9527	9521	9516	1	2	3	4	4
18	9511	9505	9500	9494	9489	9483	9478	9472	9466	9461	1	2	3	4	5
19	9455	9449	9444	9438	9432	9426	9421	9415	9409	9403	1	2	3	4	5
20	9397	9391	9385	9379	9373	9367	9361	9354	9348	9342	1	2	3	4	5
21	9336	9330	9323	9317	9311	9304	9298	9291	9285	9278	1	2	3	4	5
22	9272	9265	9259	9252	9245	9239	9232	9225	9219	9212	1	2	3	4	6
23	9205	9198	9191	9184	9178	9171	9164	9157	9150	9143	1	2	3	5	6
24	9135	9128	9121	9114	9107	9100	9092	9085	9078	9070	1	2	4	5	6
25	9063	9056	9048	9041	9033	9026	9018	9011	9003	8996	1	3	4	5	6
26	8988	8980	8973	8965	8957	8949	8942	8934	8926	8918	1	3	4	5	6
27	8910	8902	8894	8886	8878	8870	8862	8854	8846	8838	1	3	4	5	7
28	8829	8821	8813	8805	8796	8788	8780	8771	8763	8755	1	3	4	6	7
29	8746	8738	8729	8721	8712	8704	8695	8686	8678	8669	1	3	4	6	7
30	8660	8652	8643	8634	8625	8616	8607	8599	8590	8581	1	3	4	6	7
31	8572	8563	8554	8545	8536	8526	8517	8508	8499	8490	2	3	5	6	8
32	8480	8471	8462	8453	8443	8434	8425	8415	8406	8396	2	3	5	6	8
33	8387	8377	8368	8358	8348	8339	8329	8320	8310	8300	2	3	5	6	8
34	8290	8281	8271	8261	8251	8241	8231	8221	8211	8202	2	3	5	7	8
35	8192	8181	8171	8161	8151	8141	8131	8121	8111	8100	2	3	5	7	8
36	8090	8080	8070	8059	8049	8039	8028	8018	8007	7997	2	3	5	7	9
37	7986	7976	7965	7955	7944	7934	7923	7912	7902	7891	2	4	5	7	9
38	7880	7869	7859	7848	7837	7826	7815	7804	7793	7782	2	4	5	7	9
39	7771	7760	7749	7738	7727	7716	7705	7694	7683	7672	2	4	6	7	9
40	7660	7649	7638	7627	7615	7604	7593	7581	7570	7559	2	4	6	8	9
41	7547	7536	7524	7513	7501	7490	7478	7466	7455	7443	2	4	6	8	10
42	7431	7420	7408	7396	7385	7373	7361	7349	7337	7325	2	4	6	8	10
43	7314	7302	7290	7278	7266	7254	7242	7230	7218	7206	2	4	6	8	10
44	7193	7181	7169	7157	7145	7133	7120	7108	7096	7083	2	4	6	8	10

N.B.—Numbers in difference columns to be subtracted, not added.

	0'	6'	12'	18'	24'	30'	36'	42'	48'	54'	1	2	3	4	5
0°	.0000	0017	0035	0052	0070	0087	0105	0122	0140	0157	3	6	9	12	14
1	.0175	0192	0209	0227	0244	0262	0279	0297	0314	0332	3	6	9	12	15
2	.0349	0367	0384	0402	0419	0437	0454	0472	0489	0507	3	6	9	12	15
3	.0524	0542	0559	0577	0594	0612	0629	0647	0664	0682	3	6	9	12	15
4	.0699	0717	0734	0752	0769	0787	0805	0822	0840	0857	3	6	9	12	15
5	.0875	0892	0910	0928	0945	0963	0981	0998	1016	1033	3	6	9	12	15
6	.1051	1069	1086	1104	1122	1139	1157	1175	1192	1210	3	6	9	12	15
7	.1228	1246	1263	1281	1299	1317	1334	1352	1370	1388	3	6	9	12	15
8	.1405	1423	1441	1459	1477	1495	1512	1530	1548	1566	3	6	9	12	15
9	.1584	1602	1620	1638	1655	1673	1691	1709	1727	1745	3	6	9	12	15
10	.1763	1781	1799	1817	1835	1853	1871	1890	1908	1926	3	6	9	12	15
11	.1944	1962	1980	1998	2016	2035	2053	2071	2089	2107	3	6	9	12	15
12	.2126	2144	2162	2180	2199	2217	2235	2254	2272	2290	3	6	9	12	15
13	.2309	2327	2345	2364	2382	2401	2419	2438	2456	2475	3	6	9	12	15
14	.2493	2512	2530	2549	2568	2586	2605	2623	2642	2661	3	6	9	12	16
15	.2679	2698	2717	2736	2754	2773	2792	2811	2830	2849	3	6	9	13	16
16	.2867	2886	2905	2924	2943	2962	2981	3000	3019	3038	3	6	9	13	16
17	.3057	3076	3096	3115	3134	3153	3172	3191	3211	3230	3	6	10	13	16
18	.3249	3269	3288	3307	3327	3346	3365	3385	3404	3424	3	6	10	13	16
19	.3443	3463	3482	3502	3522	3541	3561	3581	3600	3620	3	6	10	13	17
20	.3640	3659	3679	3699	3719	3739	3759	3779	3799	3819	3	7	10	13	17
21	.3839	3859	3879	3899	3919	3939	3959	3978	4000	4020	3	7	10	13	17
22	.4040	4061	4081	4101	4122	4142	4163	4183	4204	4224	3	7	10	14	17
23	.4245	4265	4286	4307	4327	4348	4369	4390	4411	4431	3	7	10	14	17
24	.4452	4473	4494	4515	4536	4557	4578	4599	4621	4642	4	7	10	14	18
25	.4663	4684	4706	4727	4748	4770	4791	4813	4834	4856	4	7	11	14	18
26	.4877	4899	4921	4942	4964	4986	5008	5029	5051	5073	4	7	11	15	18
27	.5095	5117	5139	5161	5184	5206	5228	5250	5272	5295	4	7	11	15	18
28	.5317	5340	5362	5384	5407	5430	5452	5475	5498	5520	4	8	11	15	19
29	.5543	5566	5589	5612	5635	5658	5681	5704	5727	5750	4	8	12	15	19
30	.5774	5797	5820	5844	5867	5890	5914	5938	5961	5985	4	8	12	16	20
31	.6009	6032	6056	6080	6104	6128	6152	6176	6200	6224	4	8	12	16	20
32	.6249	6273	6297	6322	6346	6371	6395	6420	6445	6469	4	8	12	16	20
33	.6494	6519	6544	6569	6594	6619	6644	6669	6694	6720	4	8	13	17	21
34	.6745	6771	6796	6822	6847	6873	6899	6924	6950	6976	4	9	13	17	21
35	.7002	7028	7054	7080	7107	7133	7159	7186	7212	7239	4	9	13	18	22
36	.7265	7292	7319	7346	7373	7400	7427	7454	7481	7508	5	9	14	18	23
37	.7536	7563	7590	7618	7646	7673	7701	7729	7757	7785	5	9	14	18	23
38	.7813	7841	7869	7898	7926	7954	7983	8012	8040	8069	5	10	14	19	24
39	.8098	8127	8156	8185	8214	8243	8273	8302	8332	8361	5	10	15	20	24
40	.8391	8421	8451	8481	8511	8541	8571	8601	8632	8662	5	10	15	20	25
41	.8693	8724	8754	8785	8816	8847	8878	8910	8941	8972	5	10	16	21	26
42	.9004	9036	9067	9099	9131	9163	9195	9228	9260	9293	5	11	16	21	27
43	.9325	9358	9391	9424	9457	9490	9523	9556	9590	9623	6	11	17	22	28
44	.9657	9691	9725	9759	9793	9827	9861	9896	9930	9965	6	11	17	23	29

11	23	34	45	56
12	24	36	48	60
13	26	38	51	64
14	27	41	55	68
15	29	44	58	73
16	31	47	63	78
18	37	55	74	92
20	40	60	79	99
22	43	65	87	108
24	47	71	95	118
26	52	78	104	130
29	58	87	115	144

Difference - columns cease to be useful, owing to the rapidity with which the value of the tangent changes.

ANSWERS TO PROBLEMS

ANSWERS TO PROBLEMS

CHAPTER I

1. $2a + 6b + 6c - 3d.$
2. $-9a + b - 6c.$
3. $3d - z + 14b - 10a.$
4. $-3x + 6y + 4z + a.$
5. $-8b + 9a - 2c.$
6. $-8x - 6a + 4b + 11y.$
7. $2x - 2y + 28z$

CHAPTER II

1. $18\,a^2b^2.$
2. $48\,a^2b^2c^3.$
3. $90\,x^2y^2.$
4. $144\,a^8b^5c^2$
5. $abc^2.$
6. $\dfrac{a^2b^3c^2}{d}.$
7. $a^4b^5c.$
8. $a^3b^2c^7.$
9. $\dfrac{a^2c^2z}{b^4}.$
10. $\dfrac{40\,a^7}{c^4}.$
11. $\dfrac{b^2c^2}{54\,ad}.$

CHAPTER III

1. $\dfrac{9\,a^2b^3c}{4\,x}.$
2. $\dfrac{bc}{18\,d}.$
3. $\dfrac{a^4b^4c^2x}{16\,y^2}.$
4. $20x^2 + 15xy + 10\,xz.$
5. $4a + 2\,a^2b^2 - b.$
6. $a^2 - b^2.$
7. $6a^2 - ab + 5ac - 2b^2 + 6bc - 4c^2$
8. $a - b.$
9. $a^2 + 2\,ab + b^2.$
10. $\dfrac{a + b}{a - b}.$
11. $\dfrac{3\,a^2c - 3\,a^2d + 3\,ac^2 - 3\,acd}{2\,ac + 2\,ad - 2\,c^2 - 2\,cd}.$
12. $\dfrac{c^3ba}{12}.$

13. $\dfrac{8a + b^2 + 4c}{4b}$.

14. $\dfrac{4 - 12a + a^2c}{6a^2}$.

15. $\dfrac{120\,a^2c + 3\,bc - 6\,bx + 2\,bcd}{12\,bc}$.

16. $\dfrac{3\,ab - ac + 2\,b^2}{4\,ab}$.

17. $\dfrac{5\,a^2 - 2\,a - 2\,b}{5\,a^2 + 5\,ab}$.

CHAPTER IV

1. 3, 2, 5, a, a, b.

2. 3, 2, 2, 2, 2, a, a, a, a, c.

3. 3, 2, 5, x, x, y, y, y, y, z, z, z.

4. 3, 3, 2, 2, 2, 2, x, x, a, a.

5. 3, 2, 2, $\dfrac{1}{2}$, $\dfrac{1}{2}$, a, $\dfrac{1}{a}$, $\dfrac{1}{a}$, b, b, $\dfrac{1}{b}$, $\dfrac{1}{b}$, c, c, c.

6. 2, 5, $\dfrac{1}{2}$, x, $\dfrac{1}{x}$, $\dfrac{1}{x}$, y, y, $\dfrac{1}{y}$.

7. $(a - c)(2a + b)$.

8. $(3x + y)(x + c)$.

9. $(2x + 5y)(x + z)$.

10. $(a - b)(a - b)$.

11. $(2x - 3y)(2x - 3y)$.

12. $(9a + 5b)(9a + 5b)$.

13. $(4c - 6a)(4c - 6a)$.

14. x, y, $(4x^2 + 5zy - 10z)$.

15. $5b(6a + 3ac - c)$.

16. $(9xy - 5a)(9xy + 5a)$.

17. $(a^2 + 4b^2)(a + 2b)(a - 2b)$.

18. $(12x^2y + 8z)(12x^2y - 8z)$.

19. $(a^2 - 2ac + c)$ 2, 2.

20. $(4y + x)(4y + x)$.

21. $(3y + 2x)(2y - 3x)$.

22. $(4a + 5b)(a - 2b)$.

23. $(3y - 2x)(2y - 3x)$.

24. $(2a + b)(a - 3b)$.

25. $(2a + 5b)(a + 2b)$.

CHAPTER V

Square roots.

1. $4x + 3y$.

2. $2a + b$.

3. $6x + 2y$.

4. $5a - 2b$.

5. $a + b + c$.

Cube roots.

1. $2x + 3y$.

2. $x + 2y$.

3. $3a + 3b$.

CHAPTER VI

1. $x = 4\frac{2}{3}$.

2. $x = 2\frac{1}{3}$.

3. $x = 4$.

4. $x = -\frac{5}{13}$.

5. $x = \frac{5}{28}$.

6. $x = 30$.

7. $x = 6\frac{33}{168}$.

8. $x = \dfrac{9a + 9b - ay - by}{3}$.

9. $x = -\dfrac{3(a-b) + 2a^2}{2a(a-b)(a+1)}$.

10. $x = \dfrac{10(a^2 - b^2)}{2a}$.

11. $2a^2x + 2ab - ax^2 - bx - c^2x - bc + 10cx - 10b$.

12. $\dfrac{ax}{3} + bx = \dfrac{cy}{d} + \dfrac{3c}{d}$.

13. $a - b = \dfrac{c}{c+3}$.

14. $2 = \dfrac{10y}{y+2}$.

15. $5a + 3 = x + d + 3$.

16. $6ax - 5y = 5 - 10x$.

17. $15z^2 + 4x = 12 - 10y$.

18. $6a + 2d = 4$.

19. $3x - 2 = 3x^2 - y$.

20. $8x - 10cy = 20y$.

21. $\dfrac{x^2}{(c-d)(3a+b)} - \dfrac{x^2}{3(c-d)}$
$= 2a + b$.

22. $x = -\frac{1}{2}$.

23. Coat costs $28.57.
Gun costs $57.14.
Hat costs $14.29.

24. Horse costs $671.66.
Carriage costs $328.33.

25. Anne's age is 18 years.

26. 24 chairs and 14 tables.

CHAPTER VII

1. $y = 4, \; x = 2$.

2. $x = 5, \; y = 2$.

3. $x = 1, \; y = 2$.

4. $x = 5, \; y = 2, \; z = 3$.

5. $x = 3, \; y = 2, \; z = 4$.

6. $x = -15, \; y = 15$.

7. $x = -084, \; y = -10.034$.

8. $x = 5\frac{1}{22}, \; y = -\frac{3}{22}$.

9. $x = -1.1, \; y = 6.1$.

10. $x = 1\frac{3}{22}, \; y = 2\frac{5}{22}$.

CHAPTER VIII

1. $x = 2$ or $x = 1$.

2. $x = \dfrac{-2 \pm 2\sqrt{19}}{3}$.

3. $x = 2$.

4. $x = 4$ or -2.

5. $x = 3$ or 1.

6. $x = \pm 2$ or $\pm \sqrt{-6}$.

7. $x = -\dfrac{5 \pm \sqrt{305}}{14\,a}$.

8. $x = -\dfrac{a \pm \sqrt{12ab^2 + a^2}}{2\,b}$.

9. $x = -\dfrac{1 - 3\,a \pm \sqrt{51\,a^2 - 6\,a + 1}}{2\,a}$.

10. $x = +\dfrac{3\,(a + b) \pm \sqrt{8\,(a + b) + 9\,(a + b)^2}}{2}$.

11. $x = -\dfrac{5 \pm \sqrt{205}}{6}$.

12. $x = -3$.

13. $x = 4\left(2 \pm \sqrt{3}\right)$.

14. $x = -\dfrac{3}{4\,a}$.

15. $x = \dfrac{2\,ab}{a + b}$.

16. $x = -\dfrac{27 \pm \sqrt{2425}}{16}$.

17. $x = -\dfrac{3 \pm \sqrt{-7}}{2}$.

18. $x = -\dfrac{1 \pm \sqrt{-299}}{6}$.

19. $x = 63$.

20. $x = 100\,a^2 - 301\,a + 225$.

21. $x = \dfrac{a^2 \pm a\sqrt{a^2 + 4}}{2}$.

22. $x = \dfrac{-5 \pm \sqrt{5}}{6}$.

CHAPTER IX

1. $k = 50$.

2. $b = \sqrt{\frac{1}{441}}$.

3. $k = 60$.

4. $a = 192$.

5. $c = 5$.

CHAPTER X

1. 96 sq. ft.

2. 180 sq. ft.

3. 254.469 sq. ft.

4. Hypotenuse $= \sqrt{117}$ ft. long.

5. 62.832 ft. long.

6. $\sqrt{301}$ ft. long.

7. 27.6 ft. long.

8. 7957.7 miles.

9. Altitude $= 7.5$ ft.

10. Altitude $= 4$ ft.

CHAPTER XI

1. sine = .5349; cosine = .8456; tangent = .6330.
2. sine = .9888; cosine = .1495; tangent = 6.6122.
3. 25° 36'.   7. c = 600 ft.; b = 519.57 ft.
4. 79° 25'.   8. $\angle a$ = 57° 47'; c = 591.01 ft.
5. 36° 59'.   9. a = 1231 ft.; b = 217 ft.
6. 28° 54'.   10. $\angle a$ = 61° 51'; a = 467.3 ft.

CHAPTER XII

1. 3.5879.	10. 9,802,000.	19. 1,198,000.
2. 1.8667.	11. 24,860,000.	20. 18,410,000.
3. 3.9948.	12. 778,500,000.	21. 275,500.
4. 4.6155.	13. .000286.	22. .00001314.
5. 666.2.	14. .0001199.	23. 549.7.
6. 74430.	15. 32.34.	24. 4.27.
7. .2745.	16. 111.6.	25. .296.
8. .00024105.	17. .0323.	26. 46.86.
9. 2302.5.	18. .03767.	

CHAPTER XIII

Get cross-section paper and plot the following corresponding values of x and y and the result will be the line or curve as the case may be.

1. x = 0; y = − $3\frac{1}{3}$.   This is a straight line and only
   y = 0; x = 10.   two pairs of corresponding val-
   x = 22; y = 4.   ues of x and y are necessary to
   x = − 2; y = − 4.   draw it.

2. x = 0; y = 3.
   y = 0; x = $7\frac{1}{2}$.   This is also a straight line.

3. $x = 0;$ $y = -2.$

   $y = 0;$ $x = 4.$ } A straight line.

4. $x = 0;$ $y = -\frac{8}{10}.$

   $y = 0;$ $x = -2\frac{2}{3}.$ } A straight line.

5. $x = 0;$ $y = \pm 6.$

   $y = 0;$ $x = \pm 6.$

   $x = 1;$ $y = \pm \sqrt{35}.$

   $x = 2;$ $y = \pm \sqrt{32}.$

   $x = 3;$ $y = \pm \sqrt{27}.$

   $x = 4;$ $y = \pm \sqrt{20}.$

   $x = 5;$ $y = \pm \sqrt{11}.$

This is a circle with its center at the intersection of the x and y axes and with a radius of 6.

6. $y = 0;$ $x = 0.$

   $y = 2;$ $x = \pm \sqrt{32}.$

   $y = 4;$ $x = \pm 8.$

   $y = 6;$ $x = \pm \sqrt{96}.$

This is a parabola and to plot it correctly a great many corresponding values of x and y are necessary.

7. $y = 0;$ $x = \pm 4.$

   $y = \pm 1;$ $x = \pm \sqrt{17}.$

   $y = \pm 3;$ $x = \pm 5.$

   $y = \pm 5;$ $x = \pm \sqrt{41}.$

This is an hyperbola and a great many corresponding values of x and y are necessary in order to plot the curve correctly.

8. $y = 0;$ $x = \pm \sqrt{7}.$

   $x = 0;$ $y = +7 \text{ or } -3.$

   $x = 1;$ $y = 2 \pm \sqrt{22}.$

   $x = 2;$ $y = 2 \pm \sqrt{13}.$

This is an ellipse with its center at $+2$ on the y axis. A great many corresponding values of x and y are necessary to plot it correctly.

INTERSECTIONS OF CURVES

1. $x = 2\frac{4}{7}$; $y = 3\frac{1}{7}$. } This is the intersection of 2 straight lines.

2. $y = -5 \pm \sqrt{\frac{31}{2}}$; $x = 5 \pm \sqrt{\frac{31}{2}}$. } This is the intersection of a straight line and a circle.

3. The roots are here imaginary showing that the two curves do not touch at all, which can be easily shown by plotting them.

CHAPTER XIV

1. $6 x^2 \, \partial x$.

2. $24 x \, \partial x$.

3. $40 x^4 \, \partial x$.

4. $6 x \partial x + 4 \partial x = 15 x^2 \partial x$.

5. $8 y \, \partial y - 3 x_y \, \partial y$.

6. $42 y^4 x^2 \, \partial x + 56 x^3 y^3 \, \partial y$.

7. $\dfrac{2 yx \, \partial x - x^2 \, \partial y}{y^2}$.

8. $4 y \, \partial y - 4 q x_y \, \partial y$.

9. $y_z = -\dfrac{x}{2 y}$.

10. $y_z = -3 x^2$.

11. $y_z = \dfrac{x}{y}$.

12. $y_z = -\dfrac{y}{x}$.

13. $41° \, 48' \, 10''$.

14. When $x = 0$ at which time y also $= 0$.

15. $\dfrac{x^4}{2}$

16. $\dfrac{5 x^3}{3}$.

17. $5 ax^2 + \frac{4}{3} x^3 + 3 x$.

18. $-3 \cos x$.

19. $2 \sin x$.

20. 117.

21. 8.7795.

22. $10 \cosine x \, \partial x$.

23. $\cos^2 x \, dx - \sin^2 x \, dx$.

24. $\dfrac{1}{x} dx$.

25. $\dfrac{2 x^2 y \, \partial y - 2 y^2 x \, \partial x}{x^4}$.

SHORT=TITLE CATALOG

OF THE

Publications and Importations

OF

D. VAN NOSTRAND COMPANY

25 PARK PLACE

Prices marked with an asterisk (*) are NET
All bindings are in cloth unless otherwise noted

Abbott, A. V.   The Electrical Transmission of Energy.. ...8vo, *$5 00
—— A Treatise on Fuel.   (Science Series No. 9.)........16mo,   0 50
—— Testing Machines.   (Science Series No. 74.)........16mo,   0 50
Abraham, H.   Asphalt and Allied Substances......8vo, (*In Press.*)
Adam, P. Practical Bookbinding. Trans. by T. E. Maw.12mo,   *2 50
Adams, H.   Theory and Practice in Designing..........8vo,   *2 50
Adams, H. C.   Sewage of Seacoast Towns............8vo,   *2 00
Adams, J. W.   Sewers and Drains for Populous Districts....8vo,   2 50
Adler, A. A.   Theory of Engineering Drawing 8vo,   *2 00
—— Principles of Parallel Projecting-Line Drawing......8vo,   *1 00
Aikman, C. M.   Manures and the Principles of Manuring. . .8vo,   2 50
Aitken, W.   Manual of the Telephone................8vo,   *8 00
d'Albe, E. E. F.   Contemporary Chemistry............12mo,   *1 25
Alexander, J. H.   Elementary Electrical Engineering.....12mo,   2 00

Allan, W. Strength of Beams under Transverse Loads.
(Science Series No. 19.)....................16mo,   o 50
Allan, W. Theory of Arches. (Science Series No. 11.)..16mo,
Allen, H. Modern Power Gas Producer Practice and Applica-
tions.......................................12mo,   *2 50
Anderson, J. W. Prospector's Handbook..............12mo,   1 50
Andes, L. Vegetable Fats and Oils.....................8vo,   *5 00
—— Animal Fats and Oils. Trans. by C. Salter.........8vo,   *4 00
—— Drying Oils, Boiled Oil, and Solid and Liquid Driers...8vo,   *5 00
—— Iron Corrosion, Anti-fouling and Anti-corrosive Paints.
Trans. by C. Salter...............................8vo,   *4 00
—— Oil Colors and Printers' Ink. Trans. by A. Morris and
H. Robson.......................................8vo,   *2 50
—— Treatment of Paper for Special Purposes. Trans. by C.
Salter.......   ....12mo,   *2 50
Andrews, E. S. Reinforced Concrete Construction.....12mo,   *1 50
—— Theory and Design of Structures.................8vo,   3 50
—— Further Problems in the Theory and Design of Struc-
tures 8vo,   2 50
—— The Strength of Materials.......................8vo,   *4 00
Andrews, E. S., and Heywood, H. B. The Calculus for
Engineers 12mo,   1 50
Annual Reports on the Progress of Chemistry. Twelve Vol-
umes now ready. Vol. I, 1904, to Vol. XII, 1914, 8vo,
each......... ....   *2 00
Argand, M. Imaginary Quantities. Translated from the French
by A. S. Hardy. (Science Series No. 52.)...   .16mo,   o 50
Armstrong, R., and Idell, F. E. Chimneys for Furnaces and
Steam Boilers. (Science Series No. 1.).   ....16mo,   o 50
Arnold, E. Armature Windings of Direct Current Dynamos.
Trans. by F. B. DeGress.. ....-- ----8vo,   *2 00
Asch, W., and Asch, D. The Silicates in Chemistry and
Commerce 8vo,   *6 00
Ashe, S. W., and Keiley, J. D. Electric Railways. Theoreti-
cally and Practically Treated. Vol. I. Rolling Stock
12mo,   *2 50

Ashe, S. W.   Electric Railways.   Vol. II. Engineering Pre-
liminarie and Direct Current Sub-Stations......12mo,   *2 50
—— Electricity: Experimentally   and   Practically   Applied.
12mo,   *2 00
Ashley, R. H.   Chemical Calculations...............12mo,   *2 00
Atkins, W. Common Battery Telephony Simplified....12mo,   *1 25
Atkinson, A. A.   Electrical and Magnetic Calculations..8vo,   *1 50
Atkinson, J. J. Friction of Air in Mines. (Science Series
No. 14.)......................................16mo,   0 50
Atkinson, J. J., and Williams, E. H., Jr. Gases Met with in
Coal Mines.   (Science Series No. 13.)   16mo,   0 50
Atkinson, P. The Elements of Electric Lighting......12mo,   1 00
—— The Elements of Dynamic Electricity and Magnetism. 12mo,   2 00
Atkinson, P.   Power Transmitted by Electricity.   12mo,   2 00
Auchincloss, W. S.   Link and Valve Motions Simplified....8vo,   *1 50
Austin, E.   Single Phase Electric Railways.............4to,   *5 00
Austin and Cohn. Pocketbook of Radiotelegraphy......(*In Press.*)
Ayrton, H.   The Electric Arc..........................8vo,   *5 00
Bacon, F. W.   Treatise on the Richards Steam-Engine Indica-
tor.......................................12mo,   1 00
Bailey, R. D.   The Brewers' Analyst 8vo,   *5 00
Baker, A. L.   Quaternions............................8vo,   *1 25
—— Thick-Lens Optics...............................12mo,   *1 50
Baker, Benj. Pressure of Earthwork. (Science Series No. 56.)
16mo,
Baker, G. S.   Ship Form, Resistance and Screw Propulsion,
8vo,   *4 50
Baker, I. O   Levelling.   (Science Series No. 91.)........16mo,   0 50
Baker, M. N.   Potable Water. (Science Series No. 61).16mo,   0 50
—— Sewerage and Sewage Purification.   (Science Series No. 18.)
16mo,   0 50
Baker, T. T.   Telegraphic Transmission of Photographs.
12mo,   *1 25
Bale, G. R. Modern Iron Foundry Practice. Two Volumes.
12mo.
Vol. I. Foundry Equipment, Material Used..............   *2 50
Vol. II. Machine Moulding and Moulding Machines..   *1 50

Ball, J. W.   Concrete Structures in Railways.........8vo,   *2 50

Ball R. S.   Popular Guide to the Heavens..............8vo,   5 00

—— Natural Sources of Power.   (Westminster Series).....8vo,   *2 00

Ball, W. V.   Law Affecting Engineers...................8vo,   *3 50

Bankson, Lloyd.   Slide Valve Diagrams.   (Science Series No.
     108.)...16mo,   0 50

Barham, G. B.   Development of the Incandescent Electric
     Lamp 8vo   *2 00

Barker, A. F.   Textiles and Their Manufacture. (Westminster
     Series)...........   ....8vo,   2 00

Barker, A. F., and Midgley, E.   Analysis of Textile Fabrics,
                                             8vo,   3 00

Barker, A. H.   Graphic Methods of Engine Design....12mo,   *1 50

——Heating and Ventilation..........................4to,   *8 00

Barnard, J. H.   The Naval Militiaman's Guide..16mo, leather,   1 00

Barnard, Major J. G.   Rotary Motion.   (Science Series No. 90.)
                                             16mo,   0 50

Barnes, J. B.   Elements of Military Sketching.........16mo,   *0 75

Barrus, G. H.   Engine Tests............................8vo,   *4 00

Barwise, S.   The Purification of Sewage....   ....12mo,   3 50

Baterden, J. R.   Timber.   (Westminster Series)..........8vo,   *2 00

Bates, E. L., and Charlesworth, F.   Practical Mathematics and
     Geometry 12mo,

     Part I.   Preliminary and Elementary Course...........   *1 50
     Part II.   Advanced Course...........................   *1 50
——- Practical Mathematics............................12mo,   *1 50
——- Practical Geometry and Graphics.................12mo,   *2 00

Batey, J.   The Science of Works Management.......12mo,   *1 50

——- Steam Boilers and Combustion..................12mo,   *1 50

Bayonet Training Manual.............................16mo,   0 30

Beadle, C. Chapters on Papermaking. Five Volumes.12mo, each,   *2 00

Beaumont, R.   Color in Woven Design.................8vo,   *6 00

—— Finishing of Textile Fabrics........................8vo,   *4 00

—— Standard Cloths....................................8vo,   *5 00

Beaumont, W. W.   Steam-Engine Indicator.............8vo,   2 50

Bechhold, H. Colloids in Biology and Medicine. Trans. by J. G.
    Bullowa.............................   .(*In Press.*)
Beckwith, A. Pottery..............   ....8vo, paper,   o 6o
Bedell, F., and Pierce, C. A. Direct and Alternating Current
    Manual...8vo, *2 oo
Beech, F. Dyeing of Cotton Fabrics.. ....8vo, *4 oo
—— Dyeing of Woolen Fabrics.........................8vo, *3 50
Beggs, G. E. Stresses in Railway Girders and Bridges....(*In Press.*)
Begtrup, J. The Slide Valve..........................8vo, *2 oo
Bender, C. E. Continuous Bridges. (Science Series No. 26.)
              16mo, o 50
—— Proportions of Pins used in Bridges. (Science Series No. 4.)
              16mo, o 50
Bengough, G. D. Brass. (Metallurgy Series).........(*In Press.*)
Bennett, H. G. The Manufacture of Leather............8vo, *5 oo
Bernthsen, A. A Text-book of Organic Chemistry. Trans. by
    G. M'Gowan...................................12mo, *3 oo
Bersch, J. Manufacture of Mineral and Lake Pigments. Trans.
    by A. C. Wright........................ ......8vo, *5 oo
Bertin, L. E. Marine Boilers. Trans. by L. S. Robertson..8vo, 5 oo
Beveridge, J. Papermaker's Pocket Book.... ....12mo, *4 oo
Binnie, Sir A. Rainfall Reservoirs and Water Supply..8vo, *3 oo
Binns, C. F. Manual of Practical Potting.............8vo, *7 50
—— The Potter's Craft....   ....12mo, *2 oo
Birchmore, W. H. Interpretation of Gas Analysis.....12mo, *1 25
Blaine, R. G. The Calculus and Its Applications.......12mo, *1 50
Blake, W. H. Brewer's Vade Mecum..................8vo, *4 oo
Blasdale, W. C. Quantitative Chemical Analysis. (Van
    Nostrand's Textbooks)......................12mo, *2 50
Bligh, W. G. The Practical Design of Irrigation Works..8vo, *6 oo
Bloch, L. Science of Illumination. (Trans. by W. C.
    Clinton) 8vo, *2 50
Blok, A. Illumination and Artificial Lighting   12mo, *1 25
Blucher, H. Modern Industrial Chemistry. Trans. by J. P.
    Millington 8vo, *7 50
Blyth, A. W. Foods: Their Composition and Analysis...8vo, 7 50
—— Poisons: Their Effects and Detection.............8vo, 7 50

Böckmann, F. Celluloid.............................12mo, *2 50
Bodmer, G. R. Hydraulic Motors and Turbines........12mo, 5 00
Boileau, J. T. Traverse Tables.......   ....8vo, 5 00
Bonney, G. E. The Electro-plater's Handbook........12mo, 1 50
Booth, N. Guide to Ring-Spinning Frame...   12mo, *1 25
Booth, W. H. Water Softening and Treatment........8vo, *2 50
—— Superheaters and Superheating and their Control...8vo, *1 50
Bottcher, A. Cranes: Their Construction, Mechanical Equip-
        ment and Working. Trans. by A. Tolhausen....4to, *10 00
Bottler, M. Modern Bleaching Agents. Trans. by C. Salter.
                                            12mo, *2 50
Bottone, S. R. Magnetos for Automobilists...........12mo, *1 00
Boulton, S. B. Preservation of Timber. (Science Series No.
        82.) 16mo, 0 50
Bourcart, E. Insecticides, Fungicides and Weedkillers...8vo, *4 50
Bourgougnon, A. Physical Problems. (Science Series No. 113.)
                                            16mo, 0 50
Bourry, E. Treatise on Ceramic Industries. Trans. by
        A. B. Searle 8vo, *5 00
Bowie, A. J., Jr. A Practical Treatise on Hydraulic Mining.8vo, 5 00
Bowles, O. Tables of Common Rocks. (Science Series,
        No. 125) 16mo, 0 50
Bowser, E. A. Elementary Treatise on Analytic Geometry.12mo, 1 75
—— Elementary Treatise on the Differential and Integral
        Calculus 12mo, 2 25
Bowser, E. A. Elementary Treatise on Analytic Mechanics,
                                            12mo, 3 00
—— Elementary Treatise on Hydro-mechanics........12mo, 2 50
—— A Treatise on Roofs and Bridges...............12mo, *2 25
Boycott, G. W. M. Compressed Air Work and Diving..8vo, *4 00
Bragg, E. M. Marine Engine Design..............12mo, *2 00
—— Design of Marine Engines and Auxiliaries......... *3 00
Brainard, F. R. The Sextant. (Science Series No. 101.).16mo,
Brassey's Naval Annual for 1915. War Edition........8vo, 4 00
Briggs, R., and Wolff, A. R. Steam-Heating. (Science Series
        No. 67.) 16mo, 0 50

Bright, C. The Life Story of Sir Charles Tilson Bright. .8vo,   *4 50

Brislee, T. J. Introduction to the Study of Fuel. (Outlines
     of Industrial Chemistry.)...   ....8vo,   *3 00

Broadfoot, S. K. Motors Secondary Batteries. (Installation
     Manuals Series.) 12mo,   *0 75

Broughton, H. H. Electric Cranes and Hoists...........   *9 00

Brown, G. Healthy Foundations. (Science Series No. 80.).16mo,   0 50

Brown, H. Irrigation................................8vo,   *5 00

Brown, H. Rubber..................................8vo,   *2 00

Brown, W. A. Portland Cement Industry.............8vo,   3 00

Brown, Wm. N. The Art of Enamelling on Metal......12mo,   *1 00

—— Handbook on Japanning......................,.12mo,   *1 50

—— House Decorating and Painting.................12mo,   *1 50

—— History of Decorative Art......................12mo,   *1 25

—— Dipping, Burnishing, Lacquering and Bronzing Brass
     Ware 12mo,   *1 00

—— Workshop Wrinkles 8vo,   *1 00

Browne, C. L. Fitting and Erecting of Engines.......8vo,   *1 50

Browne, R. E. Water Meters. (Science Series No. 81.).16mo,   0 50

Bruce, E. M. Pure Food Tests......................12mo,   *1 25

Brunner, R. Manufacture of Lubricants, Shoe Polishes and
     Leather Dressings. Trans. by C. Salter........8vo,   *3 00

Buel, R. H. Safety Valves. (Science Series No. 21.)...16mo,   0 50

Burley, G. W. Lathes, Their Construction and Operation,
                                      12mo,   1 25

Burnside, W. Bridge Foundations.................12mo,   *1 50

Burstall, F. W. Energy Diagram for Gas. With text...8vo,   1 50

—— Diagram sold separately.............................   *1 00

Burt, W. A. Key to the Solar Compass......16mo, leather,   2 50

Buskett, E. W. Fire Assaying....................12mo,   *1 25

Butler, H. J. Motor Bodies and Chasis.................8vo,   *2 50

Byers, H. G., and Knight, H. G. Notes on Qualitative
     Analysis 8vo,   *1 50

Cain, W. Brief Course in the Calculus................12mo, *1 75
—— Elastic Arches. (Science Series No. 48.)........16mo, 0 50
—— Maximum Stresses. (Science Series No. 38.).....16mo, 0 50
—— Practical Dsigning Retaining of Walls. (Science Series
   No. 3.) 16mo, 0 50
—— Theory of Steel-concrete Arches and of Vaulted Struc-
   tures. (Science Series, No. 42)..............16mo, 0 50
—— Theory of Voussoir Arches. (Science Series No. 12.)
   16mo, 0 50
—— Symbolic Algebra. (Science Series No. 73.)......16mo, 0 50
Carpenter, F. D. Geographical Surveying. (Science Series
   No. 37.) 16mo,
Carpenter, R. C., and Diederichs, H. Internal-Combustion
   Engines 8vo, *5 00
Carter, H. A. Ramie (Rhea), China Grass............12mo, *2 00
Carter, H. R. Modern Flax, Hemp, and Jute Spinning...8vo, *3 00
—— Bleaching, Dyeing and Finishing of Fabrics... .8vo, *1 00
Cary, E. R. Solution of Railroad Problems With the Use of
   the Slide Rule..............................16mo, *1 00
Casler, M. D. Simplified Reinforced Concrete Mathematics,
   12mo, *1 00
Cathcart, W. L. Machine Design. Part I. Fastenings...8vo, *3 00
Cathcart, W. L., and Chaffee, J. I. Elements of Graphic
   Statics 8vo, *3 00
—— Short Course in Graphics.......................12mo, 1 50
Caven, R. M., and Lander, G. D. Systematic Inorganic Chem-
   istry 12mo, *2 00
Chalkley, A. P. Diesel Engines.....................8vo, *4 00
Chambers' Mathematical Tables......................8vo, 1 75
Chambers, G. F. Astronomy.........................16mo, *1 50
Chappel, E. Five Figure Mathematical Tables.........8vo, *2 00
Charnock. Mechanical Technology....................8vo, *3 00
Charpentier, P. Timber.............................8vo, *6 00
Chatley, H. Principles and Designs of Aeroplanes. (Science
   Series.) 16mo, 0 50
—— How to Use Water Power.........................12mo, *1 00
—— Gyrostatic Balancing............................8vo, *1 00

Child, C. D. Electric Arcs..........................8vo, *2 00
Christian, M. Disinfection and Disinfectants. Trans. by
　　Chas. Salter 12mo, 2 00
Christie, W. W. Boiler-waters, Scale, Corrosion, Foaming,
　　　　　　　　　　　　　　　　　　　　8vo, *3 00
———Chimney Design and Theory......................8vo, *3 00
———Furnace Draft. (Science Series, No. 123)........16mo, 0 50
———Water, Its Purification and Use in the Industries....8vo, *2 00
Church's Laboratory Guide. Rewritten by Edward Kinch.8vo, *1 50
Clapham, J. H. Woolen and Worsted Industries........8vo, 2 00
Clapperton, G. Practical Papermaking................8vo, 2 50
Clark, A. G. Motor Car Engineering.
　　Vol. I. Construction................................ *3 00
　　Vol. II. Design...................................(In Press.)
Clark, C. H. Marine Gas Engines...................12mo, *1 50
Clark, J. M. New System of Laying Out Railway Turnouts,
　　　　　　　　　　　　　　　　　　　　12mo, 1 00
Clarke, J. W., and Scott, W. Plumbing Practice.
　　Vol. I. Lead Working and Plumbers' Materials..8vo, *4 00
　　Vol. II. Sanitary Plumbing and Fittings..........(In Press.)
　　Vol. III. Practical Lead Working on Roofs. (In Press.)
Clarkson, R. B. Elementary Electrical Engineering.
　　　　　　　　　　　　　　　　　　　　(In Press.)
Clausen-Thue, W. A B C Universal Commercial Telegraphic
　　Code. Sixth Edition..........................(In Press.)
Clerk, D., and Idell, F. E. Theory of the Gas Engine.
　　(Science Series No. 62.).....................16mo, 0 50
Clevenger, S. R. Treatise on the Method of Government
　　Surveying . ...16mo, mor., 2 50
Clouth, F. Rubber, Gutta-Percha, and Balata.........8vo, *5 00
Cochran, J. Treatise on Cement Specifications.........8vo, *1 00
———Concrete and Reinforced Concrete Specifications....8vo, *2 50
Cochran, J. Inspection of Concrete Construction.......8vo, *4 00
Cocking, W. C. Calculations for Steel-Frame Structures.12mo, 2 25
Coffin, J. H. C. Navigation and Nautical Astronomy..12mo, *3 50
Colburn, Z., and Thurston, R. H. Steam Boiler Explosions.
　　(Science Series No. 2.).....................16mo, 0 50

Dana, R. T. Handbook of Construction Plant. .12mo, leather, *5 00
—— Handbook of Construction Efficiency..............(*In Press.*)
Danby, A. Natural Rock Asphalts and Bitumens.......8vo, *2 50
Davenport, C. The Book. (Westminster Series.)   8vo, *2 00
Davey, N. The Gas Turbine.........................8vo, *4 00
Davies, F. H.   Electric Power and Traction............8vo, *2 00
—— Foundations and Machinery Fixing. (Installation Manuals
        Series) 16mo, *1 00
Deerr, N. Sugar Cane. 8vo, 8 00
Deite, C. Manual of Soapmaking. Trans. by S. T. King..4to, *5 00
De la Coux, H. The Industrial Uses of Water. Trans. by A.
        Morris...8vo, *4 50
Del Mar, W. A. Electric Power Conductors.............8vo, *2 00
Denny, G. A. Deep-Level Mines of the Rand............4to, *10 00
—— Diamond Drilling for Gold......................... *5 00
De Roos, J. D. C. Linkages. (Science Series No. 47.)...16mo, 0 50
Derr, W. L. Block Signal Operation...........Oblong 12mo, *1 50
—— Maintenance of Way Engineering............(*In Preparation.*)
Desaint, A. Three Hundred Shades and How to Mix Them.
                                            8vo, 8 00
De Varona, A. Sewer Gases. (Science Series No. 55.)...16mo, 0 50
Devey, R. G. Mill and Factory Wiring. (Installation Manuals
        Series.)....................................12mo, *1 00
Dibdin, W. J. Purification of Sewage and Water.....8vo, 6 50
Dichman, C. Basic Open-Hearth Steel Process   8vo, *3 50
Dieterich, K. Analysis of Resins, Balsams, and Gum Resins
                                            8vo, *3 00
Dilworth, E. C. Steel Railway Bridges...............4to, *4 00
Dinger, Lieut. H. C. Care and Operation of Naval Machinery
                                            12mo. *2 00
Dixon, D. B. Machinist's and Steam Engineer's Practical Cal-
        culator..............................16mo, mor., 1 25
Dodge, G. F. Diagrams for Designing Reinforced Concrete
        Structures ....folio, *4 00
Dommett, W. E. Motor Car Mechanism.............12mo, *1 50
Dorr, B. F. The Surveyor's Guide and Pocket Table-book.
                                        16mo, mor., 2 00

Draper, C. H. Elementary Text-book of Light, Heat and
 Sound..................................... ....12mo,   1 00
Draper, C. H.   Heat and the Principles of Thermo-dynamics.
                                                        12mo,   2 00
Dron, R. W. Mining Formulas.......................12mo,   1 00
Dubbel, H. High Power Gas Engines.................8vo,   *5 00
Dumesny, P., and Noyer, J. Wood Products, Distillates, and
 Extracts...8vo,   *4 50
Duncan, W. G., and Penman, D. The Electrical Equipment of
 Collieries..... ....8vo,   *4 00
Dunkley, W. G. Design of Machine Elements...... 2 vols.
                                                12mo, each,   *1 50
Dunstan, A. E., and Thole, F. B. T. Textbook of Practical
 Chemistry..   ....12mo,   *1 40
Durham, H. W. Saws................................8vo,   2 50
Duthie, A. L. Decorative Glass Processes. (Westminster
 Series) 8vo,   *2 00
Dwight, H. B. Transmission Line Formulas...........8vo   *2 00
Dyson, S. S. Practical Testing of Raw Materials..........8vo,   *5 00
—— and Clarkson, S. S. Chemical Works..............8vo,   *7 50

Eccles, W. H. Wireless Telegraphy and Telephony..12mo,   *4 50
Eck, J. Light, Radiation and Illumination. Trans. by Paul
 Hogner 8vo,   *2 50
Eddy, H. T. Maximum Stresses under Concentrated Loads,
                                                8vo,   1 50
Eddy, L. C. Laboratory Manual of Alternating Currents,
                                                12mo,   0 50
Edelman, P. Inventions and Patents................12mo   *1 50
Edgcumbe, K. Industrial Electrical Measuring Instruments.
                                        8vo, (*In Press.*)
Edler, R. Switches and Switchgear. Trans. by Ph. Laubach.
                                                8vo,   *4 00
Eissler, M. The Metallurgy of Gold.....................8vo,   7 50
—— The Metallurgy of Silver.......................8vo,   4 00
—— The Metallurgy of Argentiferous Lead..............8vo,   5 00
—— A Handbook of Modern Explosives.................8vo,   5 00
Ekin, T. C. Water Pipe and Sewage Discharge Diagrams
                                                folio,   *3 00

Electric Light Carbons, Manufacture of...............8vo,   1 00
Eliot, C. W., and Storer, F. H.   Compendious Manual of Qualita-
    tive Chemical Analysis......................12mo,   *1 25
Ellis, C. Hydrogenation of Oils.................8vo, (*In Press.*)
Ellis, G. Modern Technical Drawing..................8vo,   *2 00
Ennis, Wm. D.   Linseed Oil and Other Seed Oils 8vo, *4 00
—— Applied Thermodynamics..........................8vo, *4 50
—— Flying Machines To-day.......................12mo, *1 50
—— Vapors for Heat Engines.......................12mo, *1 00
Ermen, W. F. A.   Materials Used in Sizing..........8vo, *2 00
Erwin, M. The Universe and the Atom.............12mo, *2 00
Evans, C. A. Macadamized Roads......................(*In Press.*)
Ewing, A. J. Magnetic Induction in Iron...............8vo, *4 00

Fairie, J. Notes on Lead Ores........................12mo, *1 00
—— Notes on Pottery Clays.........................12mo, *1 50
Fairley, W., and Andre, Geo. J. Ventilation of Coal Mines.
    (Science Series No. 58.).....................16mo,   0 50
Fairweather, W. C. Foreign and Colonial Patent Laws ...8vo, *3 00
Falk, M. S. Cement Mortars and Concretes............8vo, *2 50
Fanning, J. T. Hydraulic and Water-supply Engineering.8vo, *5 00
Fay, I. W. The Coal-tar Colors......................8vo, *4 00
Fernbach, R. L. Glue and Gelatine....................8vo, *3 00
Firth, J. B. Practical Physical Chemistry...........12mo, *1 00
Fischer, E. The Preparation of Organic Compounds. Trans.
    by R. V. Stanford.........................12mo, *1 25
Fish, J. C. L. Lettering of Working Drawings....Oblong 8vo,   1 00
—— Mathematics of the Paper Location of a Railroad,
                    12mo, paper, *0 25
Fisher, H. K. C., and Darby, W. C. Submarine Cable Testing.
                    8vo, *3 50
Fleischmann, W. The Book of the Dairy. Trans. by C. M.
    Aikman..8vo,   4 00
Fleming, J. A. The Alternate-current Transformer. Two
    Volumes......................................8vo,
    Vol. I. The Induction of Electric Currents.........   *5 00
    Vol. II. The Utilization of Induced Currents..   *5 00

Fleming, J. A. Propagation of Electric Currents......8vo, *3 00
—— A Handbook for the Electrical Laboratory and Testing
    Room. Two Volumes...................8vo, each, *5 00
Fleury, P. Preparation and Uses of White Zinc Paints..8vo, *2 50
Flynn, P. J. Flow of Water. (Science Series No. 84.).12mo, 0 50
—— Hydraulic Tables.. (Science Series No. 66.).. 16mo, 0 50
Forgie, J. Shield Tunneling...................▲...8vo. (*In Press*.)
Foster, H. A. Electrical Engineers' Pocket-book. (*Seventh
    Edition.*).............. ....12mo, leather, 5 00
—— Engineering Valuation of Public Utilities and Factories,
                                   8vo, *3 00
—— Handbook of Electrical Cost Data...........8vo. (*In Press*)
Fowle, F. F. Overhead Transmission Line Crossings....12mo, *1 50
—— The Solution of Alternating Current Problems......8vo (*In Press*.)
Fox, W. G. Transition Curves. (Science Series No. 110.).16mo, 0 50
Fox, W., and Thomas, C. W. Practical Course in Mechanical
    Drawing......................................12mo, 1 25
Foye, J. C. Chemical Problems. (Science Series No. 69.).16mo, 0 50
—— Handbook of Mineralogy. (Science Series No. 86.).
                                  16mo, 0 50
Francis, J. B. Lowell Hydraulic Experiments............4to, 15 00
Franzen, H. Exercises in Gas Analysis..............12mo, *1 00
Freudemacher, P. W. Electrical Mining Installations. (In-
    stallation Manuals Series.).................12mo, *1 00
Frith, J. Alternating Current Design.................8vo, *2 00
Fritsch, J. Manufacture of Chemical Manures. Trans. by
    D. Grant....................................8vo, *4 00
Frye, A. I. Civil Engineers' Pocket-book.......12mo, leather, *5 00
Fuller, G. W. Investigations into the Purification of the Ohio
    River....................................4to, *10 00
Furnell, J. Paints, Colors, Oils, and Varnishes.........8vo, *1 00

Gairdner, J. W. I. Earthwork.................... 8vo (*In Press*.)
Gant, L. W. Elements of Electric Traction.............8vo, *2 50
Garcia, A. J. R. V. Spanish-English Railway Terms....8vo, *4 50
Gardner, H. A. Paint Researches and Their Practical
    Application 8vo, *5 00

Garforth, W. E. Rules for Recovering Coal Mines after Explosions and Fires..........................12mo, leather, 1 50

Garrard, C. C. Electric Switch and Controlling Gear....8vo, *6 00

Gaudard, J. Foundations. (Science Series No. 34.).....16mo, 0 50

Gear, H. B., and Williams, P. F. Electric Central Station Distributing Systems.............................8vo, 3 00

Geerligs, H. C. P. Cane Sugar and Its Manufacture.......8vo, *5 00

Geikie, J. Structural and Field Geology.................8vo, *4 00

——— Mountains, Their Origin, Growth and Decay.......8vo, *4 00

——— The Antiquity of Man in Europe.................8vo, *3 00

Georgi, F., and Schubert, A. Sheet Metal Working. Trans. by C. Salter.......   ....8vo, 3 00

Gerhard, W. P. Sanitation, Water-supply and Sewage Disposal of Country Houses. 12mo, *2 00

——— Gas Lighting. (Science Series No. 111.)...........16mo, 0 50

Gerhard, W. P. Household Wastes. (Science Series No. 97.)
16mo, 0 50

——— House Drainage. (Science Series No. 63.)........16mo 0 50

——— Sanitary Drainage of Buildings. (Science Series No. 93.)
16mo, 0 50

Gerhardi, C. W. H. Electricity Meters...............8vo, *4 00

Geschwind, L. Manufacture of Alum and Sulphates. Trans. by C. Salter.................................8vo, *5 00

Gibbings, A. H. Oil Fuel Equipment for Locomotives...8vo, *2 50

Gibbs, W. E. Lighting by Acetylene................12mo, *1 50

Gibson, A. H. Hydraulics and Its Application..........8vo, *5 00

——— Water Hammer in Hydraulic Pipe Lines..........12mo, *2 00

Gibson, A. H., and Ritchie, E. G. Circular Arc Bow Girder.4to, *3 50

Gilbreth, F. B. Motion Study.......................12mo, *2 00

——— Bricklaying System.............................8vo, *3 00

——— Field System............................12mo, leather, *3 00

——— Primer of Scientific Management................12mo, *1 00

Gillette, H. P. Handbook of Cost Data.......12mo, leather, *5 00

——— Rock Excavation Methods and Cost.....12mo, leather, *5 00

——— Handbook of Earth Excavation...................(In Press.)

——— Handbook of Tunnels and Shafts, Cost and Methods of Construction...............................(In Press.)

——— Handbook of Road Construction, Methods and Cost..(In Press.)

Gunther, C. O. Integration..............................8vo, *1 25
Gurden, R. L. Traverse Tables..............folio, half mor., *7 50
Guy, A. E. Experiments on the Flexure of Beams........8vo, *1 25

Haenig, A. Emery and the Emery Industry..........8vo, *2 50
Hainbach, R. Pottery Decoration. Trans. by C. Salter.12mo, *3 00
Hale, W. J. Calculations of General Chemistry........12mo, *1 00
Hall, C. H. Chemistry of Paints and Paint Vehicles.....12mo, *2 00
Hall, G. L. Elementary Theory of Alternate Current Work-
ing 8vo, *1 50
Hall, R. H. Governors and Governing Mechanism......12mo, *2 00
Hall, W. S. Elements of the Differential and Integral Calculus
8vo, *2 25
—— Descriptive Geometry.   8vo volume and 4to atlas, *3 50
Haller, G. F., and Cunningham, E. T. The Tesla Coil.....12mo, *1 25
Halsey, F. A. Slide Valve Gears......................12mo, 1 50
—— The Use of the Slide Rule. (Science Series No. 114.)
16mo, 0 50
—— Worm and Spiral Gearing. (Science Series No. 116.)
16mo, 0 50
Hancock, H. Textbook of Mechanics and Hydrostatics.....8vo, 1 50
Hancock, W. C. Refractory Materials. (Metallurgy Series.(In Press.)
Hardy, E. Elementary Principles of Graphic Statics.....12mo, *1 50
Haring, H. Engineering Law.
Vol. I. Law of Contract 8vo, *4 00
Harper, J. H. Hydraulic Tables on the Flow of Water.16mo, *2 00
Harris, S. M. Practical Topographical Surveying. ....(In Press.)
Harrison, W. B. The Mechanics' Tool-book............12mo, 1 50
Hart, J. W. External Plumbing Work.................8vo, *3 00
—— Hints to Plumbers on Joint Wiping................8vo, *3 00
—— Principles of Hot Water Supply...................8vo, *3 00
—— Sanitary Plumbing and Drainage..................8vo, *3 00
Haskins, C. H. The Galvanometer and Its Uses........16mo, 1 50
Hatt, J. A. H. The Colorist.   ....square 12mo, *1 50
Hausbrand, E. Drying by Means of Air and Steam. Trans.
by A. C. Wright..12mo, *2 00
—— Evaporating, Condensing and Cooling Apparatus. Trans.
by A. C. Wright.......................................8vo, *5 00

Hildebrandt, A. Airships, Past and Present.............8vo, *3 50
Hildenbrand, B. W. Cable-Making. (Science Series No. 32.)
  16mo, 0 50
Hilditch, T. P. Concise History of Chemistry........12mo, *1 25
Hill, C. S. Concrete Inspection.......................16mo, *1 00
Hill, C. W. Laboratory Manual and Notes in Beginning
  Chemistry ....(*In Press.*)
Hill, J. W. The Purification of Public Water Supplies. New
  Edition.....................................(*In Press.*)
—— Interpretation of Water Analysis.....................(*In Press.*)
Hill, M. J. M. The Theory of Proportion................8vo, *2 50
Hiroi, I. Plate Girder Construction. (Science Series No. 95.)
  16mo, 0 50
—— Statically-Indeterminate Stresses..................12mo, *2 00
Hirshfeld, C. F. Engineering Thermodynamics. (Science
  Series No. 45)..............................16mo, 0 50
Hoar, A. The Submarine Torpedo Boat...............12mo, *2 00
Hobart, H. M. Heavy Electrical Engineering.............8vo, *4 50
—— Design of Static Transformers..................12mo, *2 00
—— Electricity...8vo, *2 00
—— Electric Trains......................................8vo, *2 50
—— Electric Propulsion of Ships......................8vo, *2 50
Hobart, J. F. Hard Soldering, Soft Soldering, and Brazing.
  12mo, *1 00
Hobbs, W. R. P. The Arithmetic of Electrical Measurements
  12mo, 0 50
Hoff, J. N. Paint and Varnish Facts and Formulas.....12mo, *1 50
Hole, W. The Distribution of Gas.....................8vo, *7 50
Holley, A. L. Railway Practice.......................folio, 6 00
Hopkins, N. M. Model Engines and Small Boats......12mo, 1 25
Hopkinson, J., Shoolbred, J. N., and Day, R. E. Dynamic
  Electricity. (Science Series No. 71.)..........16mo, 0 50
Horner, J. Practical Ironfounding....................8vo, *2 00
—— Gear Cutting, in Theory and Practice............8vo, *3 00
Houghton, C. E. The Elements of Mechanics of Materials.12mo, *2 00
Houstoun, R. A. Studies in Light Production.........12mo, 2 00

Induction Coils. (Science Series No. 53.)..............16mo, 0 50

Ingham, A. E. Gearing. A practical treatise..........8vo, *2 50

Ingle, H. Manual of Agricultural Chemistry. 8vo, *3 00

Inness, C. H. Problems in Machine Design............12mo, *2 00

—— Air Compressors and Blowing Engines..............12mo, *2 00

—— Centrifugal Pumps...............................12mo, *2 00

—— The Fan...12mo, *2 00

Jacob, A., and Gould, E. S. On the Designing and Construction
    of Storage Reservoirs. (Science Series No. 6.)..16mo, 0 50

Jannettaz, E. Guide to the Determination of Rocks. Trans.
    by G. W. Plympton.... ....12mo, 1 50

Jehl, F. Manufacture of Carbons......................8vo, *4 00

Jennings, A. S. Commercial Paints and Painting. (West-
    minster Series.) 8vo, *2 00

Jennison, F. H. The Manufacture of Lake Pigments · 8vo, *3 00

Jepson, G. Cams and the Principles of their Construction...8vo, *1 50

—— Mechanical Drawing.....................8vo (In Preparation.)

Jervis-Smith, F. J. Dynamometers....................8vo, *3 50

Jockin, W. Arithmetic of the Gold and Silversmith.... 12mo, *1 00

Johnson, J. H. Arc Lamps and Accessory Apparatus. (In-
    stallation Manuals Series.)..................12mo, *0 75

Johnson, T. M. Ship Wiring and Fitting. (Installation
    Manuals Series.) 12mo, *0 75

Johnson, W. McA. The Metallurgy of Nickel.........(In Preparation.)

Johnston, J. F. W., and Cameron, C. Elements of Agricultural
    Chemistry and Geology......................12mo, 2 60

Joly, J. Radioactivity and Geology..................12mo, *3 00

Jones, H. C. Electrical Nature of Matter and Radioactivity
                                          12mo, *2 00

—— Nature of Solution................................8vo, *3 50

—— New Era in Chemistry...........................12mo, *2 00

Jones, J. H. Tinplate Industry......................8vo, *3 00

Jones, M. W. Testing Raw Materials Used in Paint..... 12mo, *2 00

Jordan, L. C. Practical Railway Spiral......12mo, Leather, *1 50

Joynson, F. H. Designing and Construction of Machine Gearing................................... ....8vo, 2 00

Jüptner, H. F. V. Siderology: The Science of Iron.......8vo, *5 00

Kapp, G. Alternate Current Machinery. (Science Series No. 96.)...16mo, 0 50

Kapper, F. Overhead Transmission Lines.............4to, *4 00

Keim, A. W. Prevention of Dampness in Buildings......8vo, *2 00

Keller, S. S. Mathematics for Engineering Students.
  12mo, half leather,

—— and Knox, W. E. Analytical Geometry and Calculus.. *2 00

Kelsey, W. R. Continuous-current Dynamos and Motors.
  8vo, *2 50

Kemble, W. T., and Underhill, C. R. The Periodic Law and the Hydrogen Spectrum....................8vo, paper, *0 50

Kemp, J. F. Handbook of Rocks.......................8vo, *1 50

Kendall, E. Twelve Figure Cipher Code.................4to, *12 50

Kennedy, A. B. W., and Thurston, R. H. Kinematics of Machinery. (Science Series No. 54.)..........16mo, 0 50

Kennedy, A. B. W., Unwin, W. C., and Idell, F. E. Compressed Air. (Science Series No. 106.)...............16mo, 0 50

Kennedy, R. Electrical Installations. Five Volumes......4to, 15 00
  Single Volumes.......................................each, 3 50

—— Flying Machines; Practice and Design............12mo, *2 00

—— Principles of Aeroplane Construction..............8vo, *1 50

Kennelly, A. E. Electro-dynamic Machinery............8vo, 1 50

Kent, W. Strength of Materials. (Science Series No. 41.).16mo, 0 50

Kershaw, J. B. C. Fuel, Water and Gas Analysis........8vo, *2 50

—— Electrometallurgy. (Westminster Series.)..........8vo, *2 00

—— The Electric Furnace in Iron and Steel Production..12mo, *1 50

—— Electro-Thermal Methods of Iron and Steel Production,
  8vo, *3 00

Kindelan, J. Trackman's Helper.....................12mo, 2 00

Kinzbrunner, C. Alternate Current Windings............8vo, *1 50
—— Continuous Current Armatures....................8vo, *1 50
—— Testing of Alternating Current Machines...........8vo, *2 00
Kirkaldy, A. W., and Evans, A. D. History and Economics
    of Transport.................................8vo, *3 00
Kirkaldy. W. G. David Kirkaldy's System of Mechanical
    Testing.....................................4to, 10 00
Kirkbride. J. Engraving for Illustration...............8vo, *1 50
Kirkham, J. E. Structural Engineering...............8vo, *5 00
Kirkwood, J. P. Filtration of River Waters.............4to, 7 50
Kirschke, A. Gas and Oil Engines.. ....12mo. *1 25
Klein, J. F. Design of a High speed Steam-engine 8vo, *5 00
—— Physical Significance of Entropy. 8vo, *1 50
Klingenberg, G. Large Electric Power Stations.........4to, *5 00
Knight, R.-Adm. A. M. Modern Seamanship.. 8vo, *6 50
Knott, C. G., and Mackay, J. S. Practical Mathematics...8vo, 2 00
Knox, G. D. Spirit of the Soil.....................12mo, *1 25
Knox, J. Physico-chemical Calculations..............12mo. *1 25
—— Fixation of Atmospheric Nitrogen. (Chemical Mono-
    graphs.) ....12m^, 0 75
Koester, F. Steam-Electric Power Plants..............4to, *5 00
—— Hydroelectric Developments and Engineering........4to, *5 00
Koller, T. The Utilization of Waste Products.........8vo, *3 00
—— Cosmetics...................................8vo, *2 50
Koppe, S. W. Glycerine...........................12mo, *2 50
Kozmin, P. A. Flour Milling. Trans. by M. Falkner...8vo,
                                        (*In Press.*)
Kremann, R. Application of Physico Chemical Theory to
    Technical Processes and Manufacturing Methods.
    Trans. by H. E. Potts........................8vo, *3 00
Kretchmar, K. Yarn and Warp Sizing...............8vo, *4 00

Lallier, E. V. Elementary Manual of the Steam Engine.
                                      12mo, *2 00
Lambert, T. Lead and its Compounds.................8vo, *3 50
—— Bone Products and Manures.....................8vo, *3 00

Lamborn, L. L.   Cottonseed Products...   ....8vo,   *3 00
—— Modern Soaps, Candles, and Glycerin...............8vo,   *7 50
Lamprecht, R.   Recovery Work After Pit Fires.   Trans. by
    C. Salter ....,.............8vo,   *4 00
Lancaster, M.   Electric Cooking, Heating and Cleaning. .8vo,   *1 00
Lanchester, F. W.   Aerial Flight.   Two Volumes.   8vo.
    Vol. I.   Aerodynamics *6 00
    Vol. II.   Acrodonetics *6 00
—— The Flying Machine..............................8vo,   *3 00
Lange, K. R.   By-Products of Coal-Gas Manufacture. .12mo,   2 00
Larner, E. T.   Principles of Alternating Currents.......12mo,   *1 25
La Rue, B. F.   Swing Bridges.   (Science Series No. 107.).16mo,   0 50
Lassar-Cohn, Dr.   Modern Scientific Chemistry.   Trans. by M.
    M. Pattison Muir....................   12mo,   *2 00
Latimer, L. H., Field, C. J., and Howell, J. W.   Incandescent
    Electric Lighting.   (Science Series No. 57.).....16mo,   0 50
Latta, M. N.   Handbook of American Gas-Engineering Practice.
    8vo,   *4 50
—— American Producer Gas Practice...................4to,   *6 00
Laws, B. C.   Stability and Equilibrium of Floating Bodies.8vo,   *3 50
Lawson, W. R.   British Railways, a Financial and Commer-
    cial Survey 8vo,   2 00
Leask, A. R.   Breakdowns at Sea....................12mo,   2 00
—— Refrigerating Machinery........................12mo,   2 00
Lecky, S. T. S.   "Wrinkles" in Practical Navigation....8vo,   *10 00
Le Doux, M.   Ice Making Machines.   (Science Series No. 46.)
    16mo,   0 50
Leeds, C. C. Mechanical Drawing for Trade Schools.oblong 4to,   *2 00
—— Mechanical Drawing for High and Vocational Schools,
    4to,   *1 25
Lefévre, L.   Architectural Pottery.   Trans. by H. K. Bird and
    W. M. Binns....................................4to,   *7 50
Lehner, S.   Ink Manufacture.   Trans. by A. Morris and H
    Robson..8vo   *2 50
Lemstrom, S.   Electricity in Agriculture and Horticulture. .8vo,   *1 50
Letts, E. A.   Fundamental Problems in Chemistry   ..8vo,   *2 00

Le Van, W. B. Steam-Engine Indicator. (Science Series No. 78.)..16mo, 0 50
Lewes, V. B. Liquid and Gaseous Fuels. (Westminster Series.)
8vo, *2 00
—— Carbonization of Coal............................8vo, *3 00
Lewis Automatic Machine Rifle; Operation of..........16mo, *0 60
Lewis, L. P. Railway Signal Engineering............8vo, *3 50
Licks, H. E. Recreations in Mathematics............12mo, 1 25
Lieber, B. F. Lieber's Five Letter Standard Telegraphic Code,
8vo, *10 00
—— —— Spanish Edition 8vo, *10 00
—— —— French Edition 8vo, *10 00
—— Terminal Index 8vo, *2 50
—— Lieber's Appendix ....folio, *15 00
—— —— Handy Tables 4to, *2 50
—— Bankers and Stockbrokers' Code and Merchants and
Shippers' Blank Tables 8vo, *15 00
Lieber, B. F. 100,000,000 Combination Code.........8vo, *10 00
—— Engineering Code...   ....8vo, *12 50
Livermore, V. P., and Williams, J. How to Become a Com-
petent Motorman 12mo, *1 00
Livingstone, R. Design and Construction of Commutators.8vo, *2 25
—— Mechanical Design and Construction of Generators...8vo, *3 50
Lloyd, S. L. Fertilizer Materials......................(In Press.)
Lobben, P. Machinists' and Draftsmen's Handbook......8vo, 2 50
Lockwood, T. D. Electricity, Magnetism, and Electro-teleg-
raphy...8vo, 2 50
—— Electrical Measurement and the Galvanometer....12mo, 0 75
Lodge, O. J. Elementary Mechanics..   ....12mo, 1 50
—— Signalling Across Space without Wires..............8vo, *2 00
Loewenstein, L. C., and Crissey, C. P. Centrifugal Pumps.. *4 50
Lomax, J. W. Cotton Spinning. 12mo, 1 50
Lord, R. T. Decorative and Fancy Fabrics.............8vo, *3 50
Loring, A. E. A Handbook of the Electromagnetic Telegraph,
16mo, 0 50
—— Handbook. (Science Series No. 39).............16mo, 0 50
Lovell, D. H. Practical Switchwork. Revised by Strong and
Whitney ....(In Press.)

Low, D. A. Applied Mechanics (Elementary)........16mo, 0 80

Lubschez, B. J. Perspective.....................12mo, *1 50

Lucke, C. E. Gas Engine Design...................8vo, *3 00

—— Power Plants: their Design, Efficiency, and Power Costs.
    2 vols.................................(*In Preparation*.)

Luckiesh, M. Color and Its Application..............8vo, *3 00

——Light and Shade and Their Applications.........8vo, *2 50

    The two when purchased together................. *5 00

Lunge, G. Coal-tar Ammonia. Three Parts............8vo, *20 00

—— Manufacture of Sulphuric Acid and Alkali. Four Volumes.
                                              8vo,

    Vol. I. Sulphuric Acid. In three parts................. *18 00

    Vol. I. Supplement.................................... 5 00

    Vol. II. Salt Cake, Hydrochloric Acid and Leblanc Soda.
    In two parts.. *15 00

    Vol. III. Ammonia Soda............................. *10 00

    Vol. IV. Electrolytic Methods.....................(*In Press*.)

—— Technical Chemists' Handbook...........12mo, leather, *3 50

—— Technical Methods of Chemical Analysis. Trans. by
    C. A. Keane. In collaboration with the corps of
    specialists.

    Vol. I. In two parts.................... ....8vo, *15 00

    Vol. II. In two parts..........................8vo, *18 00

    Vol. III. In two parts..........................8vo, *18 00

    The set (3 vols.) complete.................... *50 00

—— Technical Gas Analysis...........................8vo, *4 00

Luquer, L. M. Minerals in Rock Sections..............8vo, *1 50

Macewen, H. A. Food Inspection......................8vo, *2 50

Mackenzie, N. F. Notes on Irrigation Works............8vo, *2 50

Mackie, J. How to Make a Woolen Mill Pay...........8vo, *2 00

Maguire, Wm. R. Domestic Sanitary Drainage and Plumbing
                                              8vo, 4 00

Malcolm, C. W. Textbook on Graphic Statics 8vo, *3 00

Malcolm, H. W. Submarine Telegraph Cable.........(*In Press.*)
Mallet, A. Compound Engines. Trans. by R. R. Buel.
    (Science Series No. 10.)......................16mo,
Mansfield, A. N. Electro-magnets. (Science Series No. 64)
                                 16mo,   0 50
Marks, E. C. R. Construction of Cranes and Lifting Machinery
                                 12mo, *1 50
—— Construction and Working of Pumps..............12mo, *1 50
—— Manufacture of Iron and Steel Tubes..............12mo, *2 00
—— Mechanical Engineering Materials..............12mo, *1 00
Marks, G. C. Hydraulic Power Engineering.............8vo,   3 50
—— Inventions, Patents and Designs.................12mo, *1 00
Marlow, T. G. Drying Machinery and Practice..........8vo, *5 00
Marsh, C. F. Concise Treatise on Reinforced Concrete....8vo, *2 50
Marsh, C. F. Reinforced Concrete Compression Member
    Diagram Mounted on Cloth Boards................. *1 50
Marsh, C. F., and Dunn, W. Manual of Reinforced Concrete
    and Concrete Block Construction.   .16mo, mor., *2 50
Marshall, W. J., and Sankey, H. R. Gas Engines. (Westminster
    Series.)..8vo, *2 00
Martin, G. Triumphs and Wonders of Modern Chemistry.
                                  8vo, *2 00
—— Modern Chemistry and Its Wonders...............8vo, *2 00
Martin, N. Properties and Design of Reinforced Concrete,
                                12mo, *2 50
Martin, W. D. Hints to Engineers...................12mo,   1 50
Massie, W. W., and Underhill, C. R. Wireless Telegraphy and
    Telephony.................   ....12mo, *1 00
Mathot, R. E. Internal Combustion Engines...........8vo, *4 00
Maurice, W. Electric Blasting Apparatus and Explosives ..8vo, *3 50
—— Shot Firer's Guide..................   ....8vo, *1 50
Maxwell, J. C. Matter and Motion. (Science Series No. 36.)
                                 16mo,   0 50
Maxwell, W. H., and Brown, J. T. Encyclopedia of Municipal
    and Sanitary Engineering.......................4to, *10 00
Mayer, A. M. Lecture Notes on Physics..............8vo,   2 00
Mayer, C., and Slippy, J. C. Telephone Line Construction.8vo, *3 00

McCullough, E. Practical Surveying................12mo, *2 00
—— Engineering Work in Cities and Towns..........8vo, *3 00
—— Reinforced Concrete 12mo, *1 50
McCullough, R. S. Mechanical Theory of Heat.........8vo, 3 50
McGibbon, W. C. Indicator Diagrams for Marine Engineers,
            8vo, *3 00
—— Marine Engineers' Drawing Book...........oblong 4to, *2 50
—— Marine Engineers' Pocketbook...........12mo, leather, *4 00
McIntosh, J. G. Technology of Sugar................8vo, *5 00
—— Industrial Alcohol.............................8vo, *3 00
—— Manufacture of Varnishes and Kindred Industries.
  Three Volumes. 8vo.
  Vol. I. Oil Crushing, Refining and Boiling *3 50
  Vol. II. Varnish Materials and Oil Varnish Making....... *4 00
  Vol. III. Spirit Varnishes and Materials.............. *4 50
McKnight, J. D., and Brown, A. W. Marine Multitubular
  Boilers..................................... *1 50
McMaster, J. B. Bridge and Tunnel Centres. (Science Series
  No. 20.)...... ....16mo, 0 50
McMechen, F. L. Tests for Ores, Minerals and Metals...12mo, *1 00
McPherson, J. A. Water-works Distribution 8vo, 2 50
Meade, A. Modern Gas Works Practice..... ....8vo, *7 50
Meade, R. K. Design and Equipment of Small Chemical
  Laboratories 8vo,
Melick, C. W. Dairy Laboratory Guide................12mo, *1 25
Mensch, L. J. Reinforced Concrete Pocket Book.16mo, leather *4 00
Merck, E. Chemical Reagents: Their Purity and Tests.
  Trans. by H. E. Schenck....................8vo, 1 00
Merivale, J. H. Notes and Formulae for Mining Students,
           12mo, 1 50
Merritt, Wm. H. Field Testing for Gold and Silver.16mo, leather, 1 50
Mierzinski, S. Waterproofing of Fabrics. Trans. by A. Morris
  and H. Robson......................8vo, *2 50
Miessner, B. F. Radiodynamics....................12mo, *2 00
Miller, G. A. Determinants. (Science Series No. 105.)..16mo,
Miller, W. J. Historical Geology....................12mo, *2 00
Mills, C. N. Elementary Mechanics for Engineers.....12mo, *1 00
Milroy, M. E. W. Home Lace-making....... 12mo, *1 00

Mitchell, C. A. Mineral and Aerated Waters............8vo,   *3 00
—— and Prideaux, R. M. Fibres Used in Textile and
    Allied Industries...............................8vo,   *3 00
Mitchell, C. F. and G. A. Building Construction and Draw-
    ing. 12mo
                             Elementary Course,   *1 50
                             Advanced Course,   *2 50
Monckton, C. C. F. Radiotelegraphy. (Westminster Series.)
                                        8vo,   *2 00
Monteverde, R. D. Vest Pocket Glossary of English-Spanish,
    Spanish-English Technical Terms.   ..64mo, leather,   *1 00
Montgomery, J. H. Electric Wiring Specifications....16mo,   *1 00
Moore, E. C. S. New Tables for the Complete Solution of
    Ganguillet and Kutter's Formula.........   8vo,   *5 00
Morecroft, J. H., and Hehre, F. W. Short Course in Electrical
    Testing 8vo,   *1 50
Morgan, A. P. Wireless Telegraph Apparatus for Amateurs,
                                        12mo,   *1 50
Moses, A. J. The Characters of Crystals...   ....8vo,   *2 00
—— and Parsons, C. L. Elements of Mineralogy........8vo,   *3 00
Moss, S. A. Elements of Gas Engine Design. (Science
    Series No. 121)..............................16mo,   0 50
—— The Lay-out of Corliss Valve Gears. (Science Series
    No. 119.) 16mo,   0 50
Mulford, A. C. Boundaries and Landmarks..........12mo,   *1 00
Mullin, J. P. Modern Moulding and Pattern-making....12mo,   2 50
Munby, A. E. Chemistry and Physics of Building Materials.
    (Westminster Series.)........................8vo,   *2 00
Murphy, J. G. Practical Mining....................16mo,   1 00
Murray, J. A. Soils and Manures. (Westminster Series.).8vo,   *2 00

Nasmith, J. The Student's Cotton Spinning............8vo,   3 00
—— Recent Cotton Mill Construction. 12mo,   2 50
Neave, G. B., and Heilbron, I. M. Identification of Organic
    Compounds...... ....12mo,   *1 25

Neilson, R. M.   Aeroplane Patents....... ....8vo,   *2 00
Nerz, F.   Searchlights.   Trans. by C. Rodgers...........8vo,   *3 00
Neuberger, H., and Noalhat, H.   Technology of Petroleum.
    Trans. by J. G. McIntosh.....................8vo, *10 00
Newall, J. W.   Drawing, Sizing and Cutting Bevel-gears..8vo,   1 50
Newbiging, T.   Handbook for Gas Engineers and Managers,
                             8vo,   *6 50
Newell, F. H., and Drayer, C. E.   Engineering as a Career.
                      12mo, cloth,   *1 00
                         paper,   0 75
Nicol, G.   Ship Construction and Calculations..........8vo,   *5 00
Nipher, F. E.   Theory of Magnetic Measurements.......12mo,   1 00
Nisbet, H.   Grammar of Textile Design 8vo,   *3 00
Nolan, H.   The Telescope. (Science Series No. 51.).....16mo,   0 50
North, H. B.   Laboratory Experiments in General Chemistry
                      12mo,   *1 00
Nugent, E.   Treatise on Optics.....................12mo,   1 50

O'Connor, H.   The Gas Engineer's Pocketbook...12mo, leather,   3 50
Ohm, G. S., and Lockwood, T. D.   Galvanic Circuit.   Trans. by
      William Francis. (Science Series No. 102.)....16mo,   0 50
Olsen, J. C.   Textbook of Quantitative Chemical Analysis..8vo,   *3 50
Olsson, A.   Motor Control, in Turret Turning and Gun Elevating.
    (U. S. Navy Electrical Series, No. 1.)....12mo, paper,   *0 50
Ormsby, M. T. M.   Surveying........................12mo,   1 50
Oudin, M. A.   Standard Polyphase Apparatus and Systems ..8vo,   *3 00
Owen, D.   Recent Physical Research..................8vo,   *1 50

Pakes, W. C. C., and Nankivell, A. T.   The Science of Hygiene.
                      8vo,   *1 75
Palaz, A.   Industrial Photometry.   Trans. by G. W. Patterson,
    Jr...8vo,   *4 00
Pamely, C.   Colliery Manager's Handbook.............8vo,   *10 00
Parker, P. A. M.   The Control of Water..............8vo,   *5 00
Parr, G. D. A.   Electrical Engineering Measuring Instruments.
                      8vo,   *3 50
Parry, E. J.   Chemistry of Essential Oils and Artificial Per-
    fumes.. (In Press.)

Parry, E J. Foods and Drugs. Two Volumes..........8vo,
    Vol. I. Chemical and Microscopical Analysis of Food
       and Drugs.................................... *7.50
    Vol. II. Sale of Food and Drugs Acts................ *3 00
—— and Coste, J. H. Chemistry of Pigments.........8vo, *4 50
Parry, L. Notes on Alloys...........................8vo, *3 00
—— Metalliferous Wastes 8vo, *2 00
—— Analysis of Ashes and Alloys....................8vo, *2 00
Parry, L. A. Risk and Dangers of Various Occupations.....8vo, *3 00
Parshall, H. F., and Hobart, H. M. Armature Windings 4to, *7 50
—— Electric Railway Engineering.....................4to, *10 00
Parsons, J. L. Land Drainage........................8vo, *1 50
Parsons, S. J. Malleable Cast Iron.....................8vo, *2 50
Partington, J. R. Higher Mathematics for Chemical Students
                        12mo, *2 00
—— Textbook of Thermodynamics.....................8vo, *4 00
Passmore, A. C. Technical Terms Used in Architecture ...8vo, *3 50
Patchell, W. H. Electric Power in Mines.............8vo, *4 00
Paterson, G. W. L. Wiring Calculations..............12mo, *2 00
—— Electric Mine Signalling Installations...........12mo, *1 50
Patterson, D. The Color Printing of Carpet Yarns........8vo, *3 50
—— Color Matching on Textiles.......................8vo, *3 00
—— Textile Color Mixing............................8vo, *3 00
Paulding, C. P. Condensation of Steam in Covered and Bare
       Pipes...8vo, *2 00
—— Transmission of Heat Through Cold-storage Insulation
                        12mo, *1 00
Payne, D. W. Founders' Manual.....................8vo, *4 00
Peckham, S. F. Solid Bitumens......................8vo, *5 00
Peddie, R. A. Engineering and Metallurgical Books....12mo, *1 50
Peirce, B. System of Analytic Mechanics................4to, 10 00
—— Linnear Associative Algebra......................4to, 3 00
Pendred, V. The Railway Locomotive. (Westminster Series.)
                        8vo, *2 00
Perkin, F. M. Practical Method of Inorganic Chemistry..12mo, *1 00
—— and Jaggers, E. M. Elementary Chemistry......12mo, *1 00
Perrin, J. Atoms.................................8vo, *2 50
Perrine, F. A. C. Conductors for Electrical Distribution 8vo, *3 50

Petit, G.   White Lead and Zinc White Paints............8vo,   *1 50

Petit, R.   How to Build an Aeroplane. Trans. by T. O'B.
      Hubbard, and J. H. Ledeboer.................8vo,   *1 50

Pettit, Lieut. J. S.   Graphic Processes. (Science Series No. 76.)
      16mo,   0 50

Philbrick, P. H.   Beams and Girders. (Science Series No. 88.)
      16mo,

Phillips, J.   Gold Assaying.........................8vo,   *2 50

—— Dangerous Goods...............................8vo,   3 50

Phin, J.   Seven Follies of Science...................12mo,   *1 25

Pickworth, C. N.   The Indicator Handbook. Two Volumes
      12mo, each,   1 50

—— Logarithms for Beginners ....12mo, boards,   0 50

—— The Slide Rule...............................12mo,   1 00

Plattner's Manual of Blowpipe Analysis. Eighth Edition, re-
      vised. Trans. by H. B. Cornwall..............8vo,   *4 00

Plympton, G.W.   The Aneroid Barometer. (Science Series.).16mo,   0 50

—— How to become an Engineer. (Science Series No. 100.)
      16mo,   0 50

—— Van Nostrand's Table Book. (Science Series No. 104).
      16mo,   0 50

Pochet, M. L.   Steam Injectors. Translated from the French.
      (Science Series No. 29.)....................16mo,   0 50

Pocket Logarithms to Four Places. (Science Series.).....16mo,   0 50
      leather,   1 00

Polleyn, F.   Dressings and Finishings for Textile Fabrics.8vo,   *3 00

Pope, F. G.   Organic Chemistry......................12mo,   *2 25

Pope, F. L.   Modern Practice of the Electric Telegraph. ..8vo,   1 50

Popplewell, W. C.   Prevention of Smoke..............8vo,   *3 50

—— Strength of Materials..........................8vo,   *1 75

Porritt, B. D.   The Chemistry of Rubber. (Chemical Mono-
      graphs, No. 3.)............................12mo,   *0 75

Porter, J. R.   Helicopter Flying Machine.............12mo,   *1 25

Potts, H. E. Chemistry of the Rubber Industry. (Outlines of
      Industrial Chemistry.).......................8vo,   *2 50

Practical Compounding of Oils, Tallow and Grease........8vo,   *3 50

Pratt, K.   Boiler Draught........................12mo,   *1 25
—— High Speed Steam Engines......................8vo,   *2 00
Pray, T., Jr.   Twenty Years with the Indicator..........8vo,   2 50
—— Steam Tables and Engine Constant..................8vo,   2 00
Prelini, C.   Earth and Rock Excavation................8vo,   *3 00
—— Graphical Determination of Earth Slopes............8vo,   *2 00
—— Tunneling.   New Edition........................8vo,   *3 00
—— Dredging.   A Practical Treatise...................8vo,   *3 00
Prescott, A. B.   Organic Analysis.....................8vo,   5 00
—— and Johnson, O. C.   Qualitative Chemical Analysis.8vo,   *3 50
—— and Sullivan, E. C.   First Book in Qualitative Chemistry
                                              12mo,   *1 50
Prideaux, E. B. R.   Problems in Physical Chemistry.....8vo,   *2 00
Primrose, G. S. C.   Zinc.   (Metallurgy Series.).........(*In Press.*)
Prince, G. T.   Flow of Water........................12mo,   *2 00
Pullen, W. W. F.   Application of Graphic Methods to the Design
        of Structures................................12mo,   *2 50
—— Injectors: Theory, Construction and Working......12mo,   *1 50
—— Indicator Diagrams 8vo,   *2 50
—— Engine Testing 8vo,   *4 50
Putsch, A.   Gas and Coal-dust Firing..................8vo,   *3 00
Pynchon, T. R.   Introduction to Chemical Physics........8vo,   3 00

Rafter, G. W.   Mechanics of Ventilation.   (Science Series No.
        33.).................................. ....16mo,   0 50
—— Potable Water.   (Science Series No. 103.)..........16mo,   0 50
—— Treatment of Septic Sewage.   (Science Series No. 118.)
                                              16mo,   0 50
—— and Baker, M. N.   Sewage Disposal in the United States
                                              4to,   *6 00
Raikes, H. P.   Sewage Disposal Works..............   8vo,   *4 00
Randau, P.   Enamels and Enamelling 8vo,   *4 00
Rankine, W. J. M.   Applied Mechanics.................8vo,   5 00
—— Civil Engineering................................8vo,   6 50
—— Machinery and Millwork..........................8vo,   5 00
—— The Steam-engine and Other Prime Movers.......8vo,   5 00
—— and Bamber, E. F.   A Mechanical Textbook.......8vo,   3 50

Ransome, W. R. Freshman Mathematics..............12mo,   *1 35
Raphael, F. C. Localization of Faults in Electric Light and
    Power Mains.................................8vo,   *3 50
Rasch, E. Electric Arc Phenomena. Trans. by K. Tornberg.
                                       8vo,   *2 00
Rathbone, R. L. B. Simple Jewellery...................8vo,   *2 00
Rateau, A. Flow of Steam through Nozzles and Orifices.
    Trans. by H. B. Brydon.......................8vo,   *1 50
Rausenberger, F. The Theory of the Recoil of Guns.....8vo,   *4 50
Rautenstrauch, W. Notes on the Elements of Machine Design,
                                 8vo, boards,   *1 50
Rautenstrauch, W., and Williams, J. T. Machine Drafting and
    Empirical Design.
    Part I. Machine Drafting.........................8vo,   *1 25
    Part II. Empirical Design.................... (*In Preparation.*)
Raymond, E. B. Alternating Current Engineering.....12mo,   *2 50
Rayner, H. Silk Throwing and Waste Silk Spinning...8vo,   *2 50
Recipes for the Color, Paint, Varnish, Oil, Soap and Drysaltery
    Trades 8vo,   *3 50
Recipes for Flint Glass Making.....................12mo,   *4 50
Redfern, J. B., and Savin, J. Bells, Telephones. (Installa-
    tion Manuals Series.).......................16mo,   *0 50
Redgrove, H. S. Experimental Mensuration..........12mo,   *1 25
Redwood, B. Petroleum. (Science Series No. 92.)....16mo,   0 50
Reed, S. Turbines Applied to Marine Propulsion...........   *5 00
Reed's Engineers' Handbook...........................8vo,   *6 00
—— Key to the Nineteenth Edition of Reed's Engineers'
    Handbook 8vo,   *3 00
—— Useful Hints to Sea-going Engineers.............12mo,   1 50
Reid, E. E. Introduction to Research in Organic Chemistry.
                                            (*In Press.*)
Reid, H. A. Concrete and Reinforced Concrete Construction,
                                      8vo,   *5 00
Reinhardt, C. W. Lettering for Draftsmen, Engineers, and
    Students......................oblong 4to, boards,   1 00
——The Technic of Mechanical Drafting..oblong 4to, boards,   *1 00

Reiser, F. Hardening and Tempering of Steel. Trans. by A.
        Morris and H. Robson...................12mo,   *2 50
Reiser, N. Faults in the Manufacture of Woolen Goods. Trans.
        by A. Morris and H. Robson 8vo,   *2 50
—— Spinning and Weaving Calculations.................8vo,   *5 00
Renwick, W. G. Marble and Marble Working...........8vo,   5 00
Reuleaux, F. The Constructor. Trans. by H. H. Suplee..4to,   *4 00
Reuterdahl, A. Theory and Design of Reinforced Concrete
        Arches 8vo,   *2 00
Rey, J. Range of Electric Searchlight Projectors......8vo,   *4 50
Reynolds, O., and Idell, F. E. Triple Expansion Engines.
        (Science Series No. 99.)....·.................16mo,   0 50
Rhead, G. F. Simple Structural Woodwork..........12mo,   *1 00
Rhodes, H. J. Art of Lithography.....................8vo,   3 50
Rice, J. M., and Johnson, W. W. A New Method of Obtaining
        the Differential of Functions.................12mo,   0 50
Richards, W. A. Forging of Iron and Steel.........12mo,   1 50
Richards, W. A., and North, H. B. Manual of Cement Testing,
                                                12mo,   *1 50
Richardson, J. The Modern Steam Engine............8vo,   *3 50
Richardson, S. S. Magnetism and Electricity.........12mo,   *2 00
Rideal, S. Glue and Glue Testing.....................8vo,   *4 00
Rimmer, E. J. Boiler Explosions, Collapses and Mishaps.8vo,   *1 75
Rings, F. Concrete in Theory and Practice..   12mo,   *2 50
—— Reinforced Concrete Bridges.......................4to,   *5 00
Ripper, W. Course of Instruction in Machine Drawing...folio,   *6 00
Roberts, F. C. Figure of the Earth. (Science Series No. 79.)
                                                16mo,   0 50
Roberts, J., Jr. Laboratory Work in Electrical Engineering
                                                8vo,   *2 00
Robertson, L. S. Water-tube Boilers..................8vo,   2 00
Robinson, J. B. Architectural Composition............8vo,   *2 50
Robinson, S. W. Practical Treatise on the Teeth of Wheels.
        (Science Series No. 24.)....................16mo,   0 50
—— Railroad Economics. (Science Series No. 59.)....16mo,   0 50
—— Wrought Iron Bridge Members. (Science Series No.
        60.)..16mo,   0 50

Robson, J. H. Machine Drawing and Sketching. ..8vo, *1 50

Roebling, J. A. Long and Short Span Railway Bridges.. folio, 25 00

Rogers, A. A Laboratory Guide of Industrial Chemistry..(*In Press.*)

—— Elements of Industrial Chemistry.......... 12mo, 3 00

—— Manual of Industrial Chemistry...................8vo, *5 00

Rogers, F. Magnetism of Iron Vessels. (Science Series No. 30.)
16mo, 0 50

Rohland, P. Colloidal and its Crystalloidal State of Matter.
Trans. by W. J. Britland and H. E. Potts.....12mo, *1 25

Rollinson, C. Alphabets... ....oblong 12mo, *1 00

Rose, J. The Pattern-makers' Assistant.................8vo, 2 50

—— Key to Engines and Engine-running.............. 12mo, 2 50

Rose, T. K. The Precious Metals. (Westminster Series.)..8vo, *2 00

Rosenhain, W. Glass Manufacture. (Westminster Series.)..8vo, *2 00

—— Physical Metallurgy, An Introduction to. (Metallurgy
Series.) 8vo, *3 50

Roth, W. A. Physical Chemistry..... ....8vo, *2 00

Rowan, F. J. Practical Physics of the Modern Steam-boiler.8vo, *3 00

—— and Idell, F. E. Boiler Incrustation and Corrosion.
(Science Series No. 27.)...... ....16mo, 0 50

Roxburgh, W. General Foundry Practice. (Westminster
Series.) 8vo, *2 00

Ruhmer, E. Wireless Telephony. Trans. by J. Erskine-
Murray......................................8vo, *3 50

Russell, A. Theory of Electric Cables and Networks......8vo, *3 00

Rutley, F. Elements of Mineralogy...................12mo, *1 25

Sanford, P. G. Nitro-explosives.........................8vo, *4 00

Saunders, C. H. Handbook of Practical Mechanics......16mo, 1 00
leather, 1 25

Sayers, H. M. Brakes for Tram Cars...................8vo, *1 25

Scheele, C. W. Chemical Essays.........................8vo, *2 00

Scheithauer, W. Shale Oils and Tars..................8vo, *3 50

Scherer, R. Casein. Trans. by C. Salter..............8vo, *3 00

Schidrowitz, P. Rubber, Its Production and Industrial Uses,
8vo, *5 00

Schindler, K. Iron and Steel Construction Works.....12mo, *1 25

Schmall, C. N. First Course in Analytic Geometry, Plane and
Solid...............................12mo, half leather, *1 75

Schmeer, L. Flow of Water..........................8vo, *3 00

Schumann, F. A Manual of Heating and Ventilation.
12mo, leather, 1 50

Schwartz, E. H. L. Causal Geology.............. ..8vo, *2 50

Schweizer, V. Distillation of Resins...................8vo, *4 50

Scott, W. W. Qualitative Analysis. A Laboratory Manual,
8vo, *1 50

——Standard Methods of Chemical Analysis............8vo, *6 00

Scribner, J. M. Engineers' and Mechanics' Companion.
16mo, leather, 1 50

Scudder, H. Electrical Conductivity and Ionization Constants
of Organic Compounds.........................8vo, *3 00

Searle, A. B. Modern Brickmaking..................8vo, *5 00

—— Cement, Concrete and Bricks......................8vo, *3 00

Searle, G. M. "Sumners' Method." Condensed and Improved.
(Science Series No. 124.)...................16mo, 0 50

Seaton, A. E. Manual of Marine Engineering   ..8vo, 8 00

Seaton, A. E., and Rounthwaite, H. M. Pocket-book of Marine
Engineering ....16mo, leather, 3 50

Seeligmann, T., Torrilhon, G. L., and Falconnet, H. India
Rubber and Gutta Percha. Trans. by J. G. McIntosh
8vo, *5 00

Seidell, A. Solubilities of Inorganic and Organic Substances,
8vo, *3 00

Seligman, R. Aluminum. (Metallurgy Series).........(*In Press.*)

Sellew, W. H. Steel Rails.........................4to, *10 00

—— Railway Maintenance Engineering..............12mo, *2 50

Senter, G. Outlines of Physical Chemistry...........12mo, *1 75

—— Textbook of Inorganic Chemistry...............12mo, *1 75

Sever, G. F. Electric Engineering Experiments .... 8vo, boards, *1 00

——and Townsend, F. Laboratory and Factory Tests in Electrical Engineering............   ....8vo, *2 50

Sewall, C. H.   Wireless Telegraphy...................8vo,   *2 00
—— Lessons in Telegraphy........................12mo,   *1 00
Sewell, T.   The Construction of Dynamos............8vo,   *3 00
Sexton, A. H.   Fuel and Refractory Materials .   12mo,   *2 50
—— Chemistry of the Materials of Engineering. .   12mo,   *2 50
—— Alloys (Non-Ferrous)............. .... 8vo,   *3 00
—— and Primrose, J. S. G.   The Metallurgy of Iron and Steel,
                                                  8vo,   *6 50
Seymour, A.   Modern Printing Inks..................8vo,   *2 00
Shaw, Henry S. H.   Mechanical Integrators.   (Science Series
      No. 83.) 16mo,   0 50
Shaw, S.   History of the Staffordshire Potteries.......8vo,   2 00
——Chemistry of Compounds Used in Porcelain Manufacture.8vo, *5 00
Shaw, T. R.   Driving of Machine Tools.....   12mo,   *2 00
Shaw, W. N.   Forecasting Weather...............   8vo,   *3 50
Sheldon, S., and Hausmann, E. Direct Current Machines.12mo, *2 50
—— Alternating-current Machines 12mo,   *2 50
—— Electric Traction and Transmission Engineering. .12mo,   *2 50
—— Physical Laboratory Experiments.................8vo,   *1 25
Shields, J. E.   Note on Engineering Construction..... .12mo,   1 50
Shreve, S. H.   Strength of Bridges and Roofs 8vo,   3 50
Shunk, W. F.   The Field Engineer............... 12mo, mor.,   2 50
Simmons, W. H., and Appleton, H. A.   Handbook of Soap
      Manufacture....................................8vo,   *3 00
Simmons, W. H., and Mitchell, C. A.   Edible Fats and Oils,
                                                  8vo,   *3 00
Simpson, G.   The Naval Constructor............12mo, mor.,   *5 00
Simpson, W.   Foundations........................8vo (*In Press.*)
Sinclair, A.   Development of the Locomotive Engine.
                                   8vo, half leather,   5 00
Sindall, R. W.   Manufacture of Paper.   (Westminster Series.)
                                                  8vo,   *2 00
—— and Bacon, W. N.   The Testing of Wood Pulp....8vo,   *2 50
Sloane, T. O'C.   Elementary Electrical Calculations 12mo,   *2 00
Smallwood, J. C.   Mechanical Laboratory Methods.   (Van
      Nostrand's Textbooks.)..............12mo, leather,   *2 50
Smith, C. A. M.   Handbook of Testing, MATERIALS..8vo,   *2 50
—— and Warren, A. G.   New Steam Tables............8vo,   *1 25
Smith, C. F.   Practical Alternating Currents and Testing. .8vo,   *2 50
—— Practical Testing of Dynamos and Motors..   .8vo,   *2 00

Smith, F. A. Railway Curves......................12mo,   *1 00
—— Standard Turnouts on American Railroads........12mo,   *1 00
—— Maintenance of Way Standards..................12mo,   *1 50
Smith, F. E. Handbook of General Instruction for Mechanics.
                                          12mo,   1 50
Smith, H. G. Minerals and the Microscope.........12mo,   *1 25
Smith, J. C. Manufacture of Paint...................8vo, *3 50
Smith, R. H. Principles of Machine Work..........12mo,
—— Advanced Machine Work........................12mo,   *3 00
Smith, W. Chemistry of Hat Manufacturing...........12mo, *3 00
Snell, A. T. Electric Motive Power.....................8vo, *4 00
Snow, W. G. Pocketbook of Steam Heating and Ventilation,
                                                (*In Press.*)
Snow, W. G., and Nolan, T. Ventilation of Buildings. (Science
    Series No. 5.)...............................16mo,   0 50
Soddy, F. Radioactivity.............................8vo,   *3 00
Solomon, M. Electric Lamps. (Westminster Series.).....8vo, *2 00
Somerscales, A. N. Mechanics for Marine Engineers..12mo,   *2 00
—— Mechanical and Marine Engineering Science.   .8vo,   *5 00
Sothern, J. W. The Marine Steam Turbine..   .8vo,   *6 00
—— Verbal Notes and Sketches for Marine Engineers....8vo, *7 50
Sothern, J. W., and Sothern, R. M. Elementary Mathematics
    for Marine Engineers.......................12mo,   *1 50
—— Simple Problems in Marine Engineering Design..12mo,   *1 50
Southcombe, J. E. Chemistry of the Oil Industries. (Out-
    lines of Industrial Chemistry) 8vo,   *3 00
Soxhlet, D. H. Dyeing -and Staining Marble. Trans. by A.
    Morris and H. Robson..... ....8vo,   *2 50
Spangenburg, L. Fatigue of Metals. Translated by S. H.
    Shreve. (Science Series No. 23.)..............16mo,   0 50
Specht, G. J., Hardy, A. S., McMaster, J. B., and Walling. Topo-
    graphical Surveying. (Science Series No. 72.). .16mo,   0 50
Spencer, A. S. Design of Steel-Framed Sheds.........8vo,   *3 50
Speyers, C. L. Text-book of Physical Chemistry.   ..8vo,   *1 50
Spiegel, L. Chemical Constitution and Physiological Action.
    (Trans. by C. Luedeking and A. C. Boylston.) .12mo,   *1 25

prague, E. H. Elementary Mathematics for Engineers12mo, 1 50
—— Elements of Graphic Statics....................8vo, 2 00
Sprague, E. H. Hydraulics..........................12mo, *1 50
—— Stability of Masonry.............. ....12mo, *1 50
Stahl, A. W. Transmission of Power. (Science Series No. 28.)
16mo,
—— and Woods. A. T. Elementary Mechanism......12mo, *2 00
Staley, C., and Pierson, G. S. The Separate System of
Sewerage 8vo, *3 00
Standage, H. C. Leatherworkers' Manual.............8vo, *3 50
—— Sealing Waxes, Wafers, and Other Adhesives......8vo, *2 00
—— Agglutinants of All Kinds for All Purposes......12mo, *3 50
Stanley, H. Practical Applied Physics................(In Press.)
Stansbie, J. H. Iron and Steel. (Westminster Series.)..8vo, *2 00
Steadman, F. M. Unit Photography... ....12mo, *2 00
Stecher, G. E. Cork. Its Origin and Industrial Uses..12mo, 1 00
Steinman, D. B. Suspension Bridges and Cantilevers.
(Science Series No. 127.)........................ 0 50
—— Melan's Steel Arches and Suspension Bridges......8vo, *3 00
Stevens, H. P. Paper Mill Chemist...................16mo, *2 50
Stevens, J. S. Theory of Measurements...............12mo, *1 25
Stevenson, J. L. Blast-Furnace Calculations..12mo, leather, *2 00
Stewart, G. Modern Steam Traps.............. ..:.12mo, *1 25
Stiles, A. Tables for Field Engineers........... ...12mo, 1 00
Stodola, A. Steam Turbines. Trans by L. C. Loewenstein.8vo, *5 00
Stone, H. The Timbers of Commerce.................8vo, 3 50
Stopes, M. Ancient Plants.........................8vo, *2 00
—— The Study of Plant Life........................8vo, *2 00
Sudborough, J. J., and James, T. C. Practical Organic Chem-
istry 12mo, *2 00
Suffling, E. R. Treatise on the Art of Glass Painting....8vo, *3 50
Sullivan, T. V., and Underwood, N. Testing and Valuation
of Building and Engineering Materials........(In Press.)
Sur, F. J. S. Oil Prospecting and Extracting..........8vo, *1 00
Svenson, C. L. Handbook on Piping.. ....(In Press.)
Swan, K. Patents, Designs and Trade Marks. (Westminster
Series.) 8vo, *2 00

Swinburne, J., Wordingham, C. H., and Martin, T. C.   Electric
    Currents.   (Science Series No. 109.)...........16mo,   0 50
Swoope, C. W.   Lessons in Practical Electricity........12mo,   *2 00

Tailfer, L.   Bleaching Linen and Cotton Yarn and Fabrics.8vo,   *6 00
Tate, J. S.   Surcharged and Different Forms of Retaining-
    walls.   (Science Series No. 7.)...............16mo,   0 50
Taylor, F. N.   Small Water Supplies................12mo,   *2 50
—— Masonry in Civil Engineering......................8vo,   *2 50
Taylor, T. U.   Surveyor's Handbook.........12mo, leather,   *2 00
—— Backbone of Perspective..........................12mo,   *1 00
Taylor, W. P.   Practical Cement Testing.............8vo,   *3 00
Templeton, W.   Practical Mechanic's Workshop Companion,
                         12mo, morocco,   2 00
Tenney, E. H.   Test Methods for Steam Power Plants.
    (Van Nostrand's Textbooks.)................12mo,   *2 50
Terry, H. L.   India Rubber and its Manufacture.   (West-
    minster Series.). 8vo,   *2 00
Thayer, H. R.   Structural Design.....................8vo,
    Vol.   I.   Elements of Structural Design..........   *2 00
    Vol.   II.   Design of Simple Structures.............   *4 00
    Vol.   III.   Design of Advanced Structures......(*In Preparation.*)
—— Foundations and Masonry..................(*In Preparation.*)
Thiess, J. B., and Joy, G. A.   Toll Telephone Practice..8vo,   *3 50
Thom, C., and Jones, W. H.   Telegraphic Connections,
                          oblong 12mo,   1 50
Thomas, C. W.   Paper-makers' Handbook.............(*In Press.*)
Thompson, A. B.   Oil Fields of Russia...................4to,   *7 50
—— Oil Field Development and Petroleum Mining.....8vo,   *7 50
Thompson, S. P.   Dynamo Electric Machines.   (Science
    Series No. 75.)...............   ....16mo,   0 50
Thompson, W. P.   Handbook of Patent Law of All Countries,
                                  16mo,   1 50
Thomson, G.   Modern Sanitary Engineering...........12mo,   *3 00
Thomson, G. S.   Milk and Cream Testing.............12mo,   *1 75
—— Modern Sanitary Engineering, House Drainage, etc..8vo,   *3 00

Thornley, T. Cotton Combing Machines...............8vo,   *3 00

—— Cotton Waste 8vo,   *3 00

—— Cotton Spinning 8vo,

    First Year *1 50

    Second Year *3 00

    Third Year *2 50

Thurso, J. W. Modern Turbine Practice...............8vo,   *4 00

Tidy, C. Meymott. Treatment of Sewage. (Science Series

    No. 94.) 16mo,   0 50

Tillmans, J. Water Purification and Sewage Disposal. Trans.

    by Hugh S. Taylor............................8vo,   *2 00

Tinney, W. H. Gold-mining Machinery................8vo,   *3 00

Titherley, A. W. Laboratory Course of Organic Chemistry.8vo,   *2 00

Tizard, H. T. Indicators..............................(*In Press.*)

Toch, M. Chemistry and Technology of Paints.........8vo,   *4 00

—— Materials for Permanent Painting... ....12mo,   *2 00

Tod, J., and McGibbon, W. C. Marine Engineers' Board of

    Trade Examinations 8vo,   *2 00

Todd, J., and Whall, W. B. Practical Seamanship......8vo,   8 00

Tonge, J. Coal. (Westminster Series.)...............8vo,   *2 00

Townsend, F. Alternating Current Engineering. .8vo, boards,   *0 75

Townsend, J. Ionization of Gases by Collision.........8vo,   *1 25

Transactions of the American Institute of Chemical Engineers.

    Eight volumes now ready. Vols. I. to IX., 1908-1916,

                                    8vo, each,   *6 00

Traverse Tables. (Science Series No. 115.)..........16mo,   0 50

                                      mor.,   1 00

Treiber, E. Foundry Machinery. Trans. by C. Salter..12mo,   *1 50

Trinks, W., and Housum, C. Shaft Governors. (Science

    Series No. 122.)................ .....16mo,   0 50

Trowbridge, D. C. Handbook for Engineering Draughtsmen.

                                     (*In Press.*)

Trowbridge, W. P. Turbine Wheels. (Science Series No. 44.)

                                   16mo,   0 50

Tucker, J. H. A Manual of Sugar Analysis............8vo,   3 50

Tunner, P. A. Treatise on Roll-turning. Trans. by J. B.

    Pearse...................8vo text and folio atlas,   10 00

Turnbull, Jr., J., and Robinson, S. W. A Treatise on the Compound Steam-engine. (Science Series No. 8.) 16mo,

Turner, H. Worsted Spinners' Handbook..............12mo, *2 00

Turrill, S. M. Elementary Course in Perspective......12mo, *1 25

Twyford, H. B. Purchasing............................8vo, *3 00

Tyrrell, H. G. Design and Construction of Mill Buildings.8vo, *4 00

—— Concrete Bridges and Culverts...........16mo, leather, *3 00

—— Artistic Bridge Design............................8vo, *3 00

Underhill, C. R. Solenoids, Electromagnets and Electromagnetic Windings 12mo, *2 00

Underwood, N., and Sullivan, T. V. Chemistry and Technology of Printing Inks......................8vo, ˣ3 00

Urquhart, J. W. Electro-plating.....................12mo, 2 00

—— Electrotyping 12mo, 2 00

Usborne, P. O. G. Design of Simple Steel Bridges......8vo, *4 00

Vacher, F. Food Inspector's Handbook..............12mo, *3 00

Van Nostrand's Chemical Annual. Third issue 1913. Leather, 12mo, *2 50

12mo, *2·50

—— Year Book of Mechanical Engineering Data........(In Press.)

Van Wagenen, T. F. Manual of Hydraulic Mining.....16mo, 1 00

Vega, Baron, Von. Logarithmic Tables...........8vo, cloth 2 00

half mor., 2 50

Vincent, C. Ammonia and its Compounds   Trans. by M. J. Salter 8vo, *2 00

Volk, C. Haulage and Winding Appliances.............8vo, *4 00

Von Georgievics, G. Chemical Technology of Textile Fibres. Trans. by C. Salter..... ....8vo, *4 50

—— Chemistry of Dyestuffs. Trans. by C. Salter.......8vo, *4 50

Vose, G. L. Graphic Method for Solving Certain Questions in Arithmetic and Algebra. (Science Series No. 16.) 16mo, 0 50

Vosmaer, A. Ozone......................................8vo, *2 50

Wabner, R. Ventilation in Mines. Trans . by C. Salter..8vo, *4 50

Wade, E. J. Secondary Batteries.....................8vo, *4 00

Wadmore, J. M. Elementary Chemical Theory.......12mo,   *1 50
Wadsworth, C. Primary Battery Ignition.............12mo,   *0 50
Wagner, E. Preserving Fruits, Vegetables, and Meat...12mo,   *2 50
Wagner, J. B. A Treatise on the Natural and Artificial ·
    Processes of Wood Seasoning.............8vo, (*In Press.*)
Waldram, P. J. Principles of Structural Mechanics...12mo,   *3 00
Walker, F. Aerial Navigation 8vo,   2 00
—— Dynamo Building. (Science Series No. 98.)......16mo,   0 50
Walker, J. Organic Chemistry for Students of Medicine.8vo,   *2 50
Walker, S. F. Steam Boilers, Engines and Turbines....8vo,   3 00
—— Refrigeration, Heating and Ventilation on Shipboard,
                                            12mo,   *2 00
—— Electricity in Mining............................8vo,   *3 50
Wallis-Tayler, A. J. Bearings and Lubrication.........8vo,   *1 50
—— Aerial or Wire Ropeways........................8vo,   *3 00
—— Sugar Machinery 12mo,   *2 00
Walsh, J. J. Chemistry and Physics of Mining and Mine
    Ventilation 12mo,   *2 00
Wanklyn, J. A. Water Analysis.....................12mo,   2 00
Wansbrough, W. D. The A B C of the Differential Calculus,
                                            12mo,   *1 50
—— Slide Valves 12mo,   *2 00
Waring, Jr., G. E. Sanitary Conditions. (Science Series
    No. 31.) 16mo,   0 50
—— Sewerage and Land Drainage.....................   *6 00
—— Modern Methods of Sewage Disposal............12mo,   2 00
—— How to Drain a House.........................12mo,   1 25
Warnes, A. R. Coal Tar Distillation.................8vo,   *3 00
Warren, F. D. Handbook on Reinforced Concrete......12mo,   *2 50
Watkins, A. Photography. (Westminster Series.)......8vo,   *2 00
Watson, E. P. Small Engines and Boilers............12mo,   1 25
Watt, A. Electro-plating and Electro-refining of Metals.8vo,   *4 50
—— Electro-metallurgy 12mo,   1 00
—— The Art of Soap-making.......................8vo,   3 00
—— Leather Manufacture 8vo,   *4 00
—— Paper Making 8vo,   3 00
Webb, H. L. Guide to the Testing of Insulated Wires and
    Cables 12mo,   1 00
Webber, W. H. Y. Town Gas. (Westminster Series.)...8vo,   *2 00

Weisbach, J. A Manual of Theoretical Mechanics.......8vo, *6 00
                                             sheep, *7 50
—— and Herrmann, G. Mechanics of Air Machinery....8vo, *3 75
Wells, M. B. Steel Bridge Designing...................8vo, "2 50
Weston, E. B. Loss of Head Due to Friction of Water in Pipes,
                                       12mo, *1 50
Wheatley, O. Ornamental Cement Work................8vo, *2 00
Whipple, S. An Elementary and Practical Treatise on Bridge
    Building 8vo, 3 00
White, C. H. Methods in Metallurgical Analysis. (Van
    Nostrand's Textbooks.) 12mo, 2 50
White, G. F. Qualitative Chemical Analysis..........12mo, *1 25
White, G. T. Toothed Gearing......................12mo, *1 25
Widmer, E. J. Observation Balloons...........12mo, (In Press.)
Wilcox, R. M. Cantilever Bridges. (Science Series No. 25.)
                                       16mo, 0 50
Wilda, H. Steam Turbines. Trans. by C. Salter....12mo, *1 50
—— Cranes and Hoists. Trans. by Chas. Salter........12mo, *1 50
Wilkinson, H. D. Submarine Cable Laying and Repairing.8vo, *6 00
Williamson, J. Surveying............................8vo, *3 00
Williamson, R. S. On the Use of the Barometer 4to, 15 00
—— Practical Tables in Meteorology and Hypsometry..4to, 2 50
Wilson, F. J., and Heilbron, I. M. Chemical Theory and Cal-
    culations 12mo, *1 00
Wilson, J. F. Essentials of Electrical Engineering......8vo, 2 50
Wimperis, H. E. Internal Combustion Engine............8vo, *3 00
—— Application of Power to Road Transport..........12mo, *1 50
—— Primer of Internal Combustion Engine.............12mo, *1 00
Winchell, N. H., and A. N. Elements of Optical Mineralogy.8vo, *3 50
Winslow, A. Stadia Surveying. (Science Series No. 77.).16mo, 0 50
Wisser, Lieut. J. P. Explosive Materials. (Science Series No.
    70.)...16mo, 0 50
Wisser, Lieut. J. P. Modern Gun Cotton. (Science Series No.
    89.)..16mo, 0 50
Wolff, C. E. Modern Locomotive Practice.............8vo, *4 20

Wood, De V. Luminiferous Aether. (Science Series No. 85.)
16mo, 0 50

Wood, J. K. Chemistry of Dyeing. (Chemical Monographs No. 2.) 12mo, *0 75

Worden, E. C. The Nitrocellulose Industry. Two vols..8vo, *10 00

—— Technology of Cellulose Esters. In 10 vols........8vo.
Vol. VIII. Cellulose Acetate............................. *5 00

Wren, H. Organometallic Compounds of Zinc and Magnesium. (Chemical Monographs No. 1.)...............12mo, *0 75

Wright, A. C. Analysis of Oils and Allied Substances..... 8vo, *3 50

—— Simple Method for Testing Painter's Materials.........8vo, *2 50

Wright, F. W. Design of a Condensing Plant.. 12mo, *1 50

Wright, H. E. Handy Book for Brewers...............,..8vo, *5 00

Wright, J. Testing, Fault Finding, etc. for Wiremen (Installation Manuals Series)........................16mo, *0 50

Wright, T. W. Elements of Mechanics..................8vo, *2 50

—— and Hayford, J. F. Adjustment of Observations.....8vo, *3 00

Wynne, W. E., and Spraragen, W. Handbook of Engineering Mathematics ....12mo, leather, *2 00

Yoder, J. H. and Wharen, G. B. Locomotive Valves and Valve Gears......................................8vo, 3 00

Young, J. E. Electrical Testing for Telegraph Engineers...8vo, *4 00

Zahner, R. Transmission of Power. (Science Series No. 40.)
16mo,

Zeidler, J., and Lustgarten, J. Electric Arc Lamps........8vo, *2 00

Zeuner, A. Technical Thermodynamics. Trans. by J. F. Klein. Two Volumes......................8vo, *8 co

Zimmer, G. F. Mechanical Handling and Storing of Materials,
4to, *12 50

Zipser, J. Textile Raw Materials. Trans. by C. Salter.....8vo, *5 00

Zur Nedden, F. Engineering Workshop Machines and Processes. Trans. by J. A. Davenport 8vo, *2 00

Made in the USA
Las Vegas, NV
01 November 2022

58520028R20118